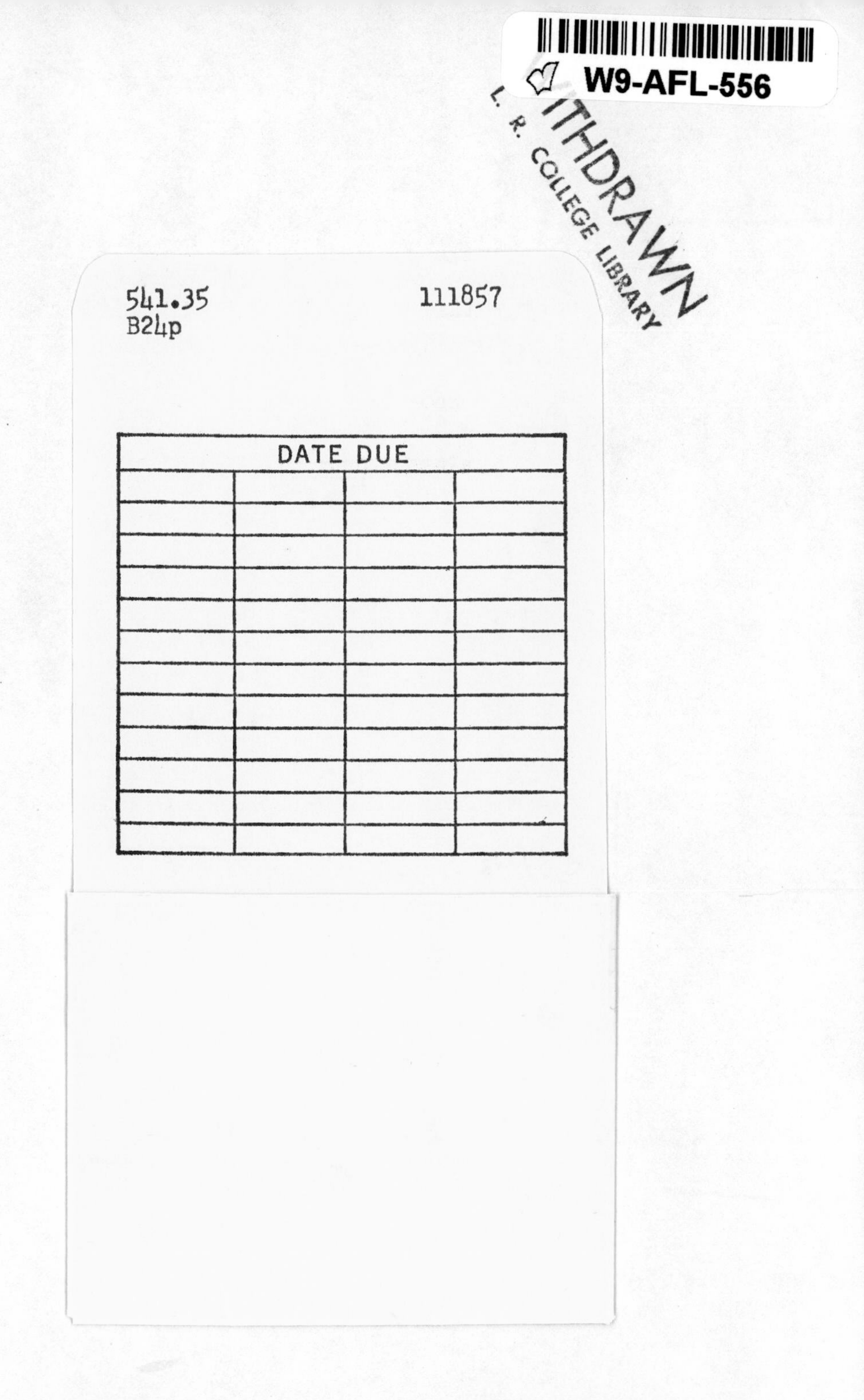

DATE DUE

PRINCIPLES OF PHOTOCHEMISTRY

J. A. Barltrop
Oxford University

and

J. D. Coyle
The Open University, Milton Keynes

JOHN WILEY & SONS

Chichester · New York · Brisbane · Toronto

Library of Congress Cataloging in Publication Data:

Barltrop, J. A.
Principles of photochemistry

Includes index.
1. Photochemistry. I. Coyle, John D., joint author.
II. Title.

QD708.2.B37 541'.35 78-16622
ISBN 0 471 99687 4

Printed in Great Britain by J. W. Arrowsmith Ltd.,
Winterstoke Road, Bristol BS3 2NT

Preface

This is a book about the electronically excited states of organic molecules—their creation and the radiative and non-radiative pathways by which they are deactivated. It is based largely on the first part of our earlier work *Excited States in Organic Chemistry*, and it arises out of a demand for a text covering the basic principles of photochemistry in a format and at a price acceptable to most students of the subject, both undergraduates and postgraduates. We have included numerous references to recent review articles and to the original literature, which we hope may be useful to postgraduates and to practising photochemists. The inclusion of a selection of problems (and answers) will, we hope, increase its appeal to undergraduates embarking on photochemistry courses.

The work is concerned with the theoretical foundations of photochemistry—with the production of excited states and with their time-independent properties and with their time-dependent behaviour. An attempt has been made to get beneath the mathematical formalism of quantum mechanics and electronic spectroscopy and to display the underlying concepts in simple (and perhaps simplistic) terms which may be readily understood and assimilated. Whether or not a particular photophysical or photochemical process is observed depends on the (often complex) relationships between its rate constant and the rate constants for alternative processes, so that kinetics are given a fundamental place in our treatment. Group theory, another powerful tool for excavations in this area, is treated in 'black box' fashion in an appendix.

We welcome any comments readers may have concerning this book.

April 1978

JOHN BARLTROP, Oxford
JOHN COYLE, Milton Keynes

Contents

Chapter 1

Introduction and Basic Principles

The field of photochemistry covers all processes which involve chemical change brought about by the action of visible or ultraviolet radiation, and these processes generally involve the direct participation of an electronically excited state of a molecule. Many life processes involve photochemical reactions, such as those of photosynthesis and vision,[1] and this reflects the fact that the major source of energy on earth is the sun's radiation. Recently this has assumed a new prominence as an intensive search is made for ways of harnessing solar energy. Photographic processes have been in use for well over a century, and these too are based on the employment of visible radiation to produce chemical change in a system.[2] Organic photochemistry, in which visible or ultraviolet radiation is used to bring about chemical reactions in organic molecules, has become a major field of study. The qualitative study of such reactions began long ago, and more detailed study of simple gas phase processes followed, but it is only since about 1950 that intensive and systematic study of liquid and solid phase photochemical processes has emerged.

With the growth of such investigations the horizons of chemistry have widened considerably. In principle, the ground electronic state of any compound can give rise to a number of different excited electronic states, each with its own characteristic properties and electron distribution, and each might have a chemistry as varied as that of the ground state. In practice, the range of observed chemical reactions of excited states is restricted by the existence of (very) rapid physical processes by which one excited state of the molecule is converted to another state of lower energy, but this still allows for an enormous amount of 'new' chemistry. Photochemical reactions have made a considerable impact in synthetic chemistry, both in research laboratories and in commercial processes. Compounds can be made by a photochemical route which are difficult, if not impossible, to prepare by a thermal method, and others can be made more readily or at lower cost. Looking to the future it seems certain that a much wider range of useful applications of photochemical reactions in synthesis will be developed, and linked with this will be the continuing search for precise mechanistic information and development of the theoretical basis.

The importance of photophysical processes in a complete description of photochemistry is the reason behind the writing of this text. The organic chemistry of photochemical processes is dealt with at a simple level in several books; here we present a complementary introduction to the photophysical processes. It is a combined study of physical and chemical processes that is leading to a greater understanding of the nature and properties of molecules in different electronic states.

1.1 THERMAL CHEMISTRY AND PHOTOCHEMISTRY

Thermal and photochemical reactions are simply different aspects of chemistry, and on the whole the same basic theoretical considerations and descriptive models can be used in both areas. The rationalization of observed chemical change in terms of electron distribution in molecules and electron re-organization during the course of a reaction step can be applied to all chemical processes, as can such secondary considerations as the effects of sterically bulky groups on the rate of reaction, or the rationalization of the stereochemical course of concerted reactions on the basis of orbital interaction. One of the major causes of difference between thermal chemistry and photochemistry lies in the differences in electron distribution in ground and excited electronic states of a molecule, which can lead to major alteration of chemical behaviour. Similarly, although thermodynamic considerations of the feasibility of reaction apply throughout chemistry, it is in this area that the cause of a second major difference between thermal chemistry and photochemistry is found. Since an electronically excited state of a molecule has a higher (often a much higher) internal energy than the ground state, there exists a much greater choice of reaction product for the excited state on thermodynamic grounds. The comparison is between a reaction $A \rightarrow B$ and a reaction $A \xrightarrow{h\nu} A^* \rightarrow B$, and there will be many systems for which $A^* \rightarrow B$ is thermodynamically favourable where the corresponding reaction $A \rightarrow B$ is not.† In particular, an excited species can give rise to high energy products such as radicals, biradicals or strained ring compounds which are not readily formed (if they can be formed at all) from the ground state.

In a photochemical reaction, thermal equilibrium between excited state, intermediate(s) and product(s) is rarely achieved, because of the magnitude of some of the energy changes involved and because of the high rate constants for many of the individual steps. It is a kinetic model of the system which is therefore often of great value in mechanistic interpretation.

For a photochemical reaction to be readily observable, the rate constant for the initial photochemical step involving the excited state must be high (typically 10^6–10^9 s^{-1}). This is because the excited state is short-lived, decaying to the ground state very rapidly, and an efficient photochemical reaction must compete successfully with these very rapid photophysical processes. The decay processes may be radiative (fluorescence or phosphorescence) or non-radiative (internal conversion or intersystem crossing). It may be that there are many photochemical reactions as yet unobserved because their quantum yields are very low as a result of such competition.

† In this book a superscript asterisk (e.g., M^*) denotes an electronically excited species. A double asterisk (M^{**}) implies an upper excited state, and where relevant a numerical superscript before the symbol ($^3M^*$ or $^1M^*$) denotes spin multiplicity as triplet or singlet respectively.

The rate constants for primary photochemical processes, like the rate constants for individual thermal reaction steps, vary with temperature, and the empirical variation can be expressed in the Arrhenius form

$$k = A\exp(-E_a/RT)$$

The activation energies for excited state reactions are generally small, often less than 30 kJ mol^{-1} (7 kcal mol^{-1}), and this is a corollary of the fact that only fast photochemical reactions are detectable.

1.2 ELECTRONIC STRUCTURE OF MOLECULES

Molecular orbitals probably afford the clearest understanding of the electronic structure of molecules and of the changes in electronic structure brought about by the absorption of electromagnetic radiation.[3,4] The molecular orbitals are formulated as linear combinations of the valence shell atomic orbitals—it is assumed that the inner electrons remain in their original atomic orbitals. For example, the interaction of two identical atomic orbitals ϕ_A and ϕ_B gives rise to two molecular orbitals of the form (equation 1.1).

$$\begin{aligned} \psi_1 &= \phi_A + \phi_B \\ \psi_2 &= \phi_A - \phi_B \end{aligned} \tag{1.1}$$

One molecular orbital is bonding (i.e. more stable than the initial atomic orbitals), and the other is antibonding (i.e. of higher energy than the initial atomic orbitals). The situation is depicted in Figure 1.1.

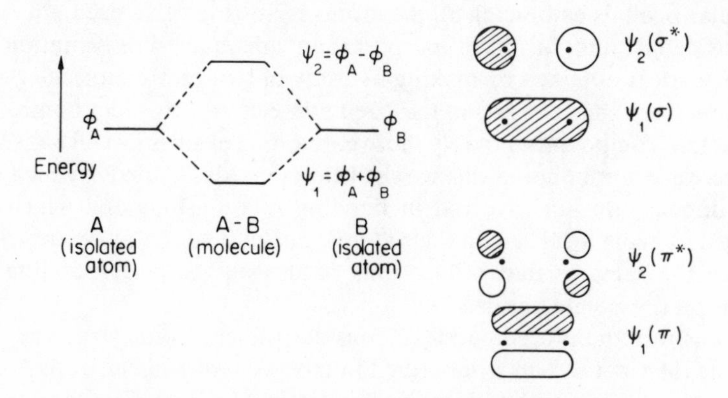

Figure 1.1. Interaction of two identical atomic orbitals

The form of the molecular orbitals is important and is also depicted in Figure 1.1. Those orbitals which are completely symmetrical about the internuclear axis are designated σ (sigma) if bonding or σ^* (sigma star) if antibonding, and these can arise if ϕ_A and ϕ_B are 's'-orbitals, for example. Molecular orbitals derived by mixing two parallel 'p'-orbitals are called π (pi) and π^* (pi star).

If the atomic orbitals are each singly occupied, or if one is doubly occupied and the other is vacant, the electrons in the molecular system both occupy the low-energy bonding molecular orbital. This leads to a gain in stability over the isolated atoms, and it is the basis of the molecular orbital description of electron-pair covalent bonding.

Molecular orbitals can encompass more than two atomic centres, and this leads to electron delocalization. The π-molecular orbitals of buta-1,3-diene are an example of this, and they are obtained by taking linear combinations of the four C(2p) orbitals. Their form is shown in Figure 1.2.

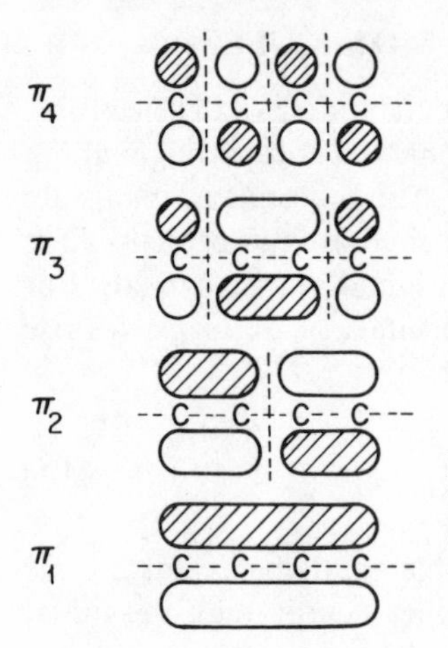

Figure 1.2. The π-molecular orbitals of buta-1,3-diene. The wavefunction has opposite signs in regions of the orbitals which are cross-hatched or left blank. Nodes, where the wavefunction changes sign, are shown as dotted lines

In a similar way to this, the σ-framework of an organic molecule consists of molecular orbitals embracing all the atoms.† However, this need not normally be considered, and for most purposes an adequate representation of the σ-framework is obtained by making use only of two-centre molecular orbitals. The framework thus consists of localized two-electron covalent bonds.

In certain compounds, notably those containing elements of Groups V, VI or VII, there are non-bonding valence shell electrons (designated *n*) which, as their name implies, are not involved in bonding relationships and which can be regarded as being localized on their atomic nuclei. The energy of such electrons is much the same as that of electrons occupying the corresponding atomic orbitals on the isolated atom.

To illustrate the preceding ideas, consider the electronic structure of formaldehyde ($H_2C{=}O$). This is described in terms of pairs of electrons occupying three localized σ-bonding molecular orbitals (C—H, C—H, C—O), one localized π-bonding molecular orbital (C—O), and two non-degenerate, non-bonding orbitals on oxygen (a p-orbital and a sp hybrid orbital).[5] The 'core' 1s electrons on carbon and oxygen are ignored. The shapes of these orbitals and the anti-bonding orbitals are shown in Figure 1.3.

† Note that the σ–π approximation is adopted here, which implies a lack of interaction between the σ- and π-orbitals because of their different symmetry. This approximation is adequate for most discussions of molecular structure.

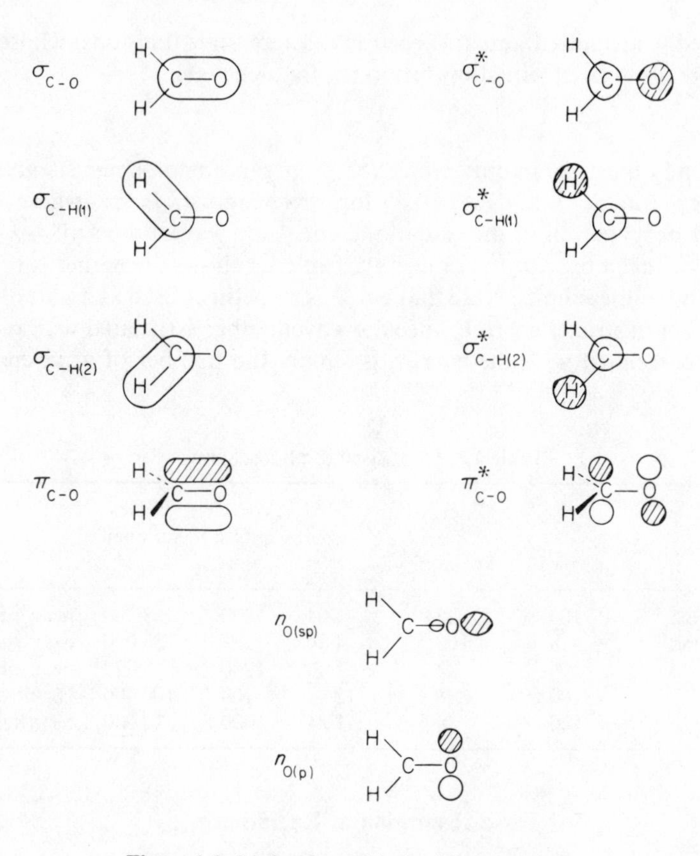

Figure 1.3. Molecular orbitals of formaldehyde

1.3 ELECTROMAGNETIC RADIATION

Electromagnetic radiation, of which visible light and ultraviolet radiation are examples, can be envisaged in terms of an oscillating electric field and an oscillating magnetic field operating in planes which are perpendicular to each other and to the direction of propagation. The time-variable strength of each field at a given point is described by a sinusoidal function. In a beam of normal radiation the orientation of the fields with respect to the surroundings is random, but plane polarized radiation, in which this orientation is restricted to a particular plane, can be produced using certain ordered arrays of ions in crystals or of molecules in a matrix to absorb and transmit selectively radiation with a particular direction of polarization. A property of such plane polarized light is that the direction of polarization is changed by passage through an ordered or a random array of chiral molecules.

For some purposes it is more convenient to use a particle description of electromagnetic radiation, since radiation of a given frequency is quantized and is emitted, transmitted and absorbed in discrete units (photons) whose energy (E) is directly related (equation 1.2) to the frequency (ν).

$$E = h\nu \quad (1.2)$$

The units most commonly employed by organic photochemists are s^{-1} for frequency, nm or Å (1 nm = 10 Å) for wavelength ($\lambda = c/\nu$, where c is the speed of propagation of the radiation), cm^{-1} for wavenumber ($\bar{\nu} = \lambda^{-1}$), and kJ mol^{-1}, kcal mol^{-1} or eV for energy. Table 1.1 shows the numerical relationship between these units. Note that cm^{-1} is sometimes used as a unit of energy, but this is not strictly correct, since the wavenumber associated with radiation, though correlated with the energy, is simply the number of wavelengths per cm.

Table 1.1. Units used in photochemistry

	Energy			Wave-length	Wave-number	
	kJ mol^{-1}	kcal mol^{-1}	eV	nm	cm^{-1}	
100 kJ mol^{-1}	100	23·9	1·04	1200	8 360	near infrared
100 kcal mol^{-1}	418	100	4·34	286	35 000	near ultraviolet
1 eV	96·5	23·1	1·00	1240	8 070	near infrared
100 nm	1200	286	12·4	100	100 000	far ultraviolet
10 000 cm^{-1}	120	28·6	1·24	1000	10 000	near infrared

1.3.1 Absorption of Radiation

When a photon passes close to a molecule there is an interaction between the electric field associated with the molecule and that associated with the radiation. This perturbation may result in no permanent change in the molecule, but it is possible for a 'reaction' to occur in which the photon is absorbed by the molecule. The photon ceases to exist and its energy is transferred to the molecule, whose electronic structure changes. This change is visualized in simple molecular orbital terms as a change in the occupation pattern of a set of orbitals which is the same set in the excited state as in the ground state. This is the one-electron excitation approximation, and the approximation holds good for visualizing most absorption processes, although in some instances (e.g. for Rydberg transitions in alkenes see Chapter 2, p. 30) it is necessary to consider orbitals not normally envisaged for a description of the ground state. Formaldehyde offers a simple example (Figure 1.4: note that only the C—O orbitals are shown). The absorption of a photon corresponding to radiation of wavelength around 280 nm produces an electronically excited state of the carbonyl group in which there is only one electron in the non-bonding orbital of higher energy, and one electron in the anti-bonding π^* orbital.

Such a transition is referred to as an $n \rightarrow \pi^*$ (n to pi star) transition, and the excited state as an (n, π^*) excited state of formaldehyde. Other types of transition

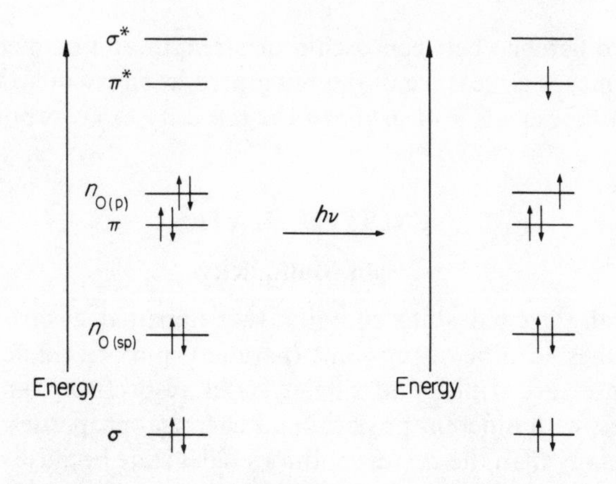

Figure 1.4. $n \rightarrow \pi^*$ Electronic excitation of formaldehyde

are possible with different wavelengths of radiation or with different classes of compound. Those most commonly encountered in organic compounds are $n \rightarrow \pi^*$, $\pi \rightarrow \pi^*$, $n \rightarrow \sigma^*$, and $\sigma \rightarrow \sigma^*$. The energy of the photon will not often match exactly the energy difference between the lowest vibrational levels of the ground and excited states of the absorbing molecule, so that in most cases the state initially produced is an upper vibrational/rotational state of the excited electronic state.

Under normal circumstances the bulk absorption characteristics of a compound (vapour, liquid, solid or solution) can be represented[6] by equation (1.3).

$$I = I_0\,10^{-\varepsilon cl}$$

or

$$\log(I_0/I) = \varepsilon cl \qquad (1.3)$$

I_0 is the intensity of the incident monochromatic radiation, I is the intensity of transmitted radiation, c is the concentration (or partial pressure or density) of the sample, l is the path-length of the radiation through the sample, and ε is a constant which is characteristic of the particular compound and the particular wavelength of radiation. ε is known as the extinction coefficient, or more specifically as the decadic molar extinction coefficient if c is in molarity units, l is in cm, and logarithms are to base 10.

This empirical law (the Beer–Lambert law) is valid except when very high intensities of radiation are employed (e.g., when using lasers) and a significant proportion of the molecules in a given region are in the excited state rather than the ground state at any one time.

An alternative measure of absorption intensity, which can be related more readily to theoretical principles,[7] is the oscillator strength, f, given by equation (1.4).

$$f = 4{\cdot}315 \times 10^{-9} \int \varepsilon \, . \, \mathrm{d}\nu \qquad (1.4)$$

The major difference between oscillator strength and extinction coefficient is that the former is a measure of the integrated intensity of absorption over a whole band, whereas ε is a measure of the intensity of absorption for a single wavelength.

1.4 EXCITED STATES

1.4.1 Spin Multiplicity

An electronically excited state contains two unpaired electrons in different orbitals, and these can be of the same (parallel) spin or of different (opposed) spin. Such states are triplet and singlet states respectively, and the two are distinct species, with different physical and chemical properties. A triplet state has a lower energy than the corresponding singlet state because of the repulsive nature of the spin–spin interaction between electrons of the same spin (Hund's rule of maximum multiplicity for atomic electronic structure has the same basis). The magnitude of the difference in energy varies according to the degree of spatial interaction (overlap) between the orbitals involved. For orbitals which occupy substantially different regions of space [as in (n, π^*) states of carbonyl compounds] orbital overlap is small and the singlet–triplet energy difference (splitting) is relatively small. For orbitals which occupy similar regions of space [as in (π, π^*) states of alkenes] the difference is much larger.

The excited states initially produced by absorption of a photon are almost always singlet states. This is because practically all molecules encountered in organic chemistry have a singlet ground state (i.e. are fully electron-paired), and the selection rules for absorption strongly favour conservation of spin during the absorption process. Singlet → triplet absorption bands in the absorption spectra of some compounds can be observed with sensitive spectrophotometers, and these bands can often be enhanced by the presence of a paramagnetic species such as molecular oxygen, but they are, in general, very much weaker than singlet → singlet bands.

1.4.2 State Diagrams

Orbital energy level diagrams have been used in this chapter to show the electronic structure of a particular state of a molecule. A different type of diagram, a state energy level diagram, can be used to represent diagrammatically the various electronic states of a molecule. The singlet and triplet states are ranged in order of increasing energy (Figure 1.5) and numbered in the same order S_0, S_1, S_2 ... and T_1, T_2 ... respectively. For clarity the states of different multiplicity are separated horizontally, and normally only the first few (lowest energy) states are of interest. Each of the energy levels is the lowest energy point of a complete potential energy surface (energy versus all the variable parameters of the molecule), and the horizontal axis in a simple state diagram has no significance.

It is possible to represent on such a diagram all the physical processes involved in the interconversion of the states (Figure 1.6). These are absorption of a photon

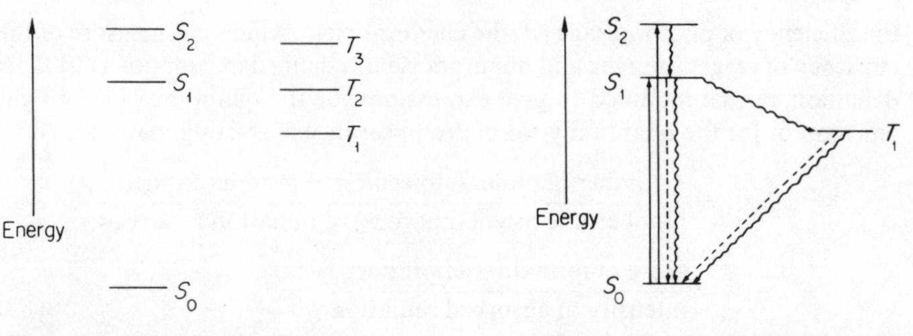

Figure 1.5. A simple state diagram

Figure 1.6. A simple Jablonski diagram

by the ground state to produce an excited singlet state (→), radiative decay processes (- - →), namely fluorescence (spin-allowed) and phosphorescence (spin-forbidden), and non-radiative decay processes (⤳), namely internal conversion (spin-allowed) and intersystem crossing (spin-forbidden). Such a diagram is known as a *Jablonski* diagram.

Other pathways open to each excited state are chemical reaction and energy transfer. The latter is a bimolecular process, often referred to as *quenching*, in which the excited state is converted to ground state in a non-radiative manner whilst the second molecule (the 'quencher' Q) is excited to a higher energy state. All the processes can be encompassed in a flow diagram (equation 1.5).

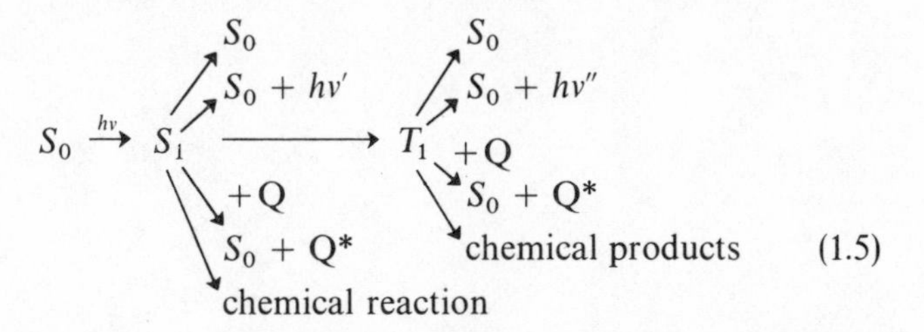

(1.5)

In the chemical reaction the process may be a simple concerted formation of products from reagents, or it may occur through one or more intermediates which may also be able to revert to ground state starting material (equation 1.6). The nature of the secondary chemical processes will depend on the system in question.

$$S_1 \text{ or } T_1 \rightarrow \text{intermediate} \begin{matrix} \nearrow S_0 \\ \searrow \text{product} \end{matrix} \qquad (1.6)$$

Such diagrams bring out the point that the overall efficiency of any particular process is governed by the relative magnitude of a number of individual rate constants. A useful parameter in quantitative photochemistry to which these rate constant ratios can be related is the quantum yield (ϕ). This is a measure of the efficiency of photon usage (cf. the chemical yield, which is a measure of the efficiency of reagent usage), and quantum yield is defined in equation (1.7). This definition can be modified to give expressions for the quantum yield of light emission or for the quantum yield of disappearance of starting material.

$$\phi_{\text{product}} = \frac{\text{number of moles (molecules) of product formed}}{\text{number of einstein (photons) of radiation absorbed}} = \frac{\text{rate of formation of product}}{\text{intensity of absorbed radiation}} \qquad (1.7)$$

REFERENCES

1. J. B. Thomas, *Primary Photoprocesses in Biology*, North-Holland, Amsterdam (1965); D. O. Cowan and R. L. Drisko, *Elements of Organic Photochemistry*, Plenum Press, New York (1976), chapter 12.
2. T. H. James (ed.), *The Theory of the Photographic Process*, 3rd edition, Macmillan, New York (1966).
3. A. Streitwieser, *Molecular Orbital Theory for Organic Chemists*, Wiley, London (1961), chapter 1.
4. H. H. Jaffé and M. Orchin, *Theory and Applications of Ultraviolet Spectroscopy*, Wiley, London (1962), chapter 3.
5. For an alternative molecular orbital description of formaldehyde, see reference 4, p. 105.
6. Reference 4, p. 8.
7. R. S. Mulliken, *J. Chem. Phys.*, 7, 14 (1939).

Chapter 2

Excited States. Production and Time-independent Properties

There are many ways of introducing energy into molecules in order to produce electronically excited states, but by far the most important involves absorption of light (usually visible or ultraviolet). This is therefore treated first; consideration of other methods of generating excited states is deferred to p. 36.

2.1 FACTORS AFFECTING INTENSITIES OF ABSORPTION SPECTRA

Analysis of absorption and emission spectra and of their variation as a function of experimental conditions provides a powerful tool for exploring the nature of excited states. An understanding of the factors determining the form and intensities of such spectra is essential, and this requires a consideration of the probabilities associated with particular transitions. The following section presents a non-mathematical theoretical treatment of absorption and emission phenomena—the latter will be studied in more detail in Chapter 3.

2.1.1 Absorption and Emission of Light

The absorption of light can be regarded as an exercise in time-dependent perturbation theory. A molecule in an initial stationary state described by the wavefunction Ψ_i is subject to the Schrödinger equation $H_0\Psi_i = E\Psi_i$. If the system is perturbed by exposing it to light, the sinusoidal oscillating electric vector of the light wave induces oscillating forces on the charged particles of the molecule. Thus the static Hamiltonian operator H_0 no longer prescribes the energy of the system, and it has to be replaced by $(H_0 + H')$, where H' is the perturbation operator which takes into account the effect of the radiation field.

The eigenfunctions of $(H_0 + H')$ will be different from the initial wavefunction Ψ_i, and they will also be functions of time. Thus:

$$(H_0 + H')\Psi(x, t) = E\Psi(x, t)$$

These new wavefunctions can be expanded in terms of the wavefunctions of the unperturbed system:

$$\Psi(x, t) = \sum a_k(t)\Psi_k$$

where the coefficients $a_k(t)$, being functions of time, bring in the required time-dependence.

Thus the effect of the perturbation can be thought of as a time-dependent mixing of the initial wavefunction of the system with all the other possible wavefunctions. In other words, the initial state evolves with time into other states, and if the perturbation is suddenly removed at a time t there will be a finite probability that the system will be found in some final state (Ψ_f) other than the initial one. This probability is given by the square of the corresponding coefficient $a_f(t)$ in the above expansion. Mathematical analysis[1] leads to the expression in equation (2.1).

$$[a_f(t)]^2 = \frac{8\pi^3}{3h^2}\langle\Psi_i|\boldsymbol{\mu}|\Psi_f\rangle^2\rho(\nu_{if})t \tag{2.1}$$

where $\rho(\nu_{if})$ is the radiation density (energy per unit volume) at the frequency ν_{if} corresponding to the transition, t is the time of irradiation, and $\langle\Psi_i|\boldsymbol{\mu}|\Psi_f\rangle$, which is merely a shorthand way of writing the integral $\int \Psi_i \,.\, \boldsymbol{\mu} \,.\, \Psi_f \,d\tau$, is an extremely important quantity known as the *transition moment* ($\boldsymbol{\mu}$ is the dipole moment operator $e\sum \boldsymbol{r}_j$, where e is the electronic charge and $\boldsymbol{r}_j$ is the distance of the jth electron).

This shows, then, that the probability of a given transition is proportional to the square of the transition moment. This quantity, which, in principle, can be calculated quantum mechanically, is also obtainable from absorption spectra through its relation (equation 2.2) to the oscillator strength f (see Chapter 1, p. 7).

$$f = \frac{8\pi^2\nu_{if}m_e\langle\Psi_i|\boldsymbol{\mu}|\Psi_f\rangle^2}{3he^2} \tag{2.2}$$

(m_e is the electronic mass.)

If the final state has higher energy than the initial state, the energy deficit must be made up from the radiation field and equation (2.1) gives the probability of *absorption* of a photon. However, equation (2.1) applies equally to the converse situation in which the initial state has higher energy than the final state. This implies that, on irradiation with light of the frequency corresponding to the transition, an excited species may be induced to revert to the ground state with the emission of a photon. This phenomenon, known as *stimulated emission*, provides the basis for laser action (see p. 41).

Absorption and emission of radiation are often discussed in terms of the Einstein coefficients which give the transition rate for absorption and emission per unit radiation density and time. The coefficient of (stimulated) absorption

B_{lu} (where the subscripts u and l refer to upper and lower states) is given by

$$B_{lu} = \frac{[a_u(t)]^2}{\rho(\nu_{lu})t} = \frac{8\pi^3 \langle \Psi_l | \boldsymbol{\mu} | \Psi_u \rangle^2}{3h^2} \quad \text{from equation (2.1)}$$

Similarly, the Einstein coefficient of stimulated emission is B_{ul}, and from the preceding discussion,

$$B_{ul} = B_{lu}$$

This being so, if there were no other effect, irradiation of atoms and molecules at the transition frequency would produce an equal population of the upper and lower states, for the probabilities of absorption and stimulated emission are the same. This is well known not to be the case.

In order to account for thermal equilibrium in a radiation field, Einstein found it necessary to assume that there was another emission process with a rate *independent* of the radiation density and given by his coefficient of *spontaneous* emission A_{ul}. Hence the overall rate of emission is equal to

$$A_{ul} + B_{ul}\rho(\nu_{ul})$$

It can be shown that A_{ul} and B_{ul} are related by the expression in equation (2.3).

$$A_{ul} = \frac{8\pi h\nu^3}{c^3} \, . \, B_{ul} = \frac{64\pi^4 \nu^3}{3hc^3} \, . \, \langle \Psi_u | \boldsymbol{\mu} | \Psi_l \rangle^2 \tag{2.3}$$

This relation codifies the important points that the rates of absorption and emission (spontaneous and stimulated) depend on the square of the transition moment, but that the rate of spontaneous emission depends also on the *cube* of the transition frequency. Thus upper excited states may be expected for this reason, if for no other, to be very short-lived.

2.1.2 Radiative Lifetimes

If no other decay process exists, a population of electronically excited species will decay radiatively to the ground state. Spontaneous emission, being random, obeys first-order kinetics with a rate constant equal to the Einstein coefficient A_{ul} which is the number of times per second that an excited state emits a photon.

$$\frac{d[M^*]}{dt} = -A_{ul}[M^*]$$
$$\therefore \quad [M^*] = M_0^* \, e^{-A_{ul}t} \tag{2.4}$$

We can therefore define a *radiative lifetime*

$$\tau_0 = \frac{1}{A_{ul}} \quad \text{(units, seconds per transition)}$$

which is the time for the population to diminish to 1/e of its initial concentration, assuming that *no radiationless processes are occurring.*

Since A_{ul} is directly related to B_{ul} and the oscillator strength f (via equations 2.2 and 2.3), it follows that when the absorption leading to a particular excited state is intense (i.e., the transition is allowed and f is large), the radiative lifetime of the excited state is short. Conversely, for a forbidden transition (e.g., singlet $\longrightarrow$ triplet) the corresponding radiative lifetime is long. In other words, if the absorption of radiation is forbidden/allowed, the emission is also forbidden/allowed.

An order-of-magnitude estimate of the radiative lifetime may be obtained from the relation (2.5) for absorption bands in the near ultraviolet.

$$\tau_0 \sim 10^{-4}/\varepsilon_{max} \tag{2.5}$$

Thus, an allowed transition with $\varepsilon \sim 10^5$ l mol^{-1} cm^{-1} will give rise to an excited state having an approximate radiative lifetime of 10^{-9} s (1 ns). Alternatively, for forbidden $S \longrightarrow T$ transitions $\varepsilon \sim 10^{-1}$–10^{-4} l mol^{-1} cm^{-1} and τ_0 can be of the order of seconds. This enormous radiative lifetime is one of the reasons why triplet species play such an important role in photochemistry. For more precise estimates of radiative lifetimes, several more complex relations have been proposed.[2]

Actual Lifetimes

It is important to make a clear distinction between the *radiative* (τ_0) and *actual* lifetimes (τ). Consider a system (2.6) in which an excited species A* both fluoresces with a rate constant $k_f (\equiv A_{ul})$ and undergoes a reaction with rate constant k_r.

$$\mathrm{A} \xrightarrow[h\nu]{I} \mathrm{A}^* \begin{cases} \xrightarrow{k_f} \mathrm{A} + h\nu' \\ \xrightarrow{k_r} \text{products} \end{cases} \tag{2.6}$$

The actual rate constant for the decay of A* is $(k_r + k_f)$, and the *actual* lifetime of A* (as measured by the decay of fluorescence) will be found to be:

$$\tau = \frac{1}{k_r + k_f}$$

Generalizing, if there are several modes of decay (radiative and non-radiative) each characterized by a first-order or pseudo-first-order rate constant k_i ($i = 1, 2, \ldots, n$), then the actual lifetime (i.e., decay time) will be:

$$\tau = \frac{1}{\sum_i k_i} \tag{2.7}$$

The two lifetimes are related by the expression:

$$\tau = \tau_0 \phi_f$$

where ϕ_f is the quantum yield of fluorescence.

2.1.3 The Intensities of Electronic Transitions

We have seen that the rates of absorption and spontaneous emission, and therefore the intensities, of electronic transitions are proportional to the square of the transition moment (T.M.) $\langle \Psi_i|\boldsymbol{\mu}|\Psi_f \rangle$, where Ψ is the *total* wavefunction (nuclear and electronic) for the system. This integral cannot be evaluated, because we do not know the exact form of the wavefunctions of any molecule. In order to make progress, approximations must be introduced.

Born–Oppenheimer Approximation

Because of the mass difference, nuclear motion is very sluggish in comparison with electronic motion, so that the electrons may be thought of as moving in the potential field of the static nuclei. If the Schrödinger equation can be solved for a variety of nuclear configurations and the electronic potential energy is plotted as a function of these nuclear configurations, we obtain a potential energy surface. The wavefunctions obtained in this way differ from the true wavefunctions, but the differences are usually important only near degeneracies, where potential surfaces cross.

This, the Born–Oppenheimer approximation, is equivalent mathematically to factorizing the total wavefunction into a nuclear (vibrational) wavefunction θ and an electronic wavefunction ψ. Thus $\Psi = \theta \, . \, \psi$ and the transition moment is given by equation (2.8).†

$$\text{T.M.} = \int \theta_i \psi_i \, . \, \boldsymbol{\mu} \, . \, \theta_f \psi_f \, d\tau \tag{2.8}$$

Since the operator $\boldsymbol{\mu}$ operates only on the electrons, we can write this in the form of equation (2.9):

$$\text{T.M.} = \int \theta_i \theta_f \, d\tau_N \, . \int \psi_i \, . \, \boldsymbol{\mu} \, . \, \psi_f \, d\tau_e \tag{2.9}$$

where subscripts N and e refer to nuclei and electrons. This expression is still too difficult, and further approximations must be made.

(a) It is assumed (i) that ψ can be represented as the product of one-electron wavefunctions (orbitals) ϕ (which may themselves be linear combinations of atomic orbitals) and that the orbitals are the same in both ground and excited states, and (ii) that only one electron is promoted during a transition. The transition moment then reduces to the expression (2.10), where ϕ_i and ϕ_f are the initial and final *orbitals* of the excited electron and $\boldsymbol{\mu}$ operates only on these orbitals.

$$\text{T.M.} = \int \theta_i \theta_f \, d\tau_N \, . \int \phi_i \, . \, \boldsymbol{\mu} \, . \, \phi_f \, d\tau_e \tag{2.10}$$

† Strictly, complex wavefunctions should be used, with T.M. $= \int \theta_i^* \psi_i^* \, . \, \boldsymbol{\mu} \, . \, \theta_f \psi_f \, d\tau$, but in this discussion it will be assumed that orbital wavefunctions are real. Complex wavefunctions are likely to be important only in systems of high symmetry such as linear polyatomics.

(b) The final approximation is that the orbitals can be factorized into a product of space and spin orbitals (φ and S respectively),

$$\phi = \varphi . S$$

Since $\boldsymbol{\mu}$ operates only on the space coordinates,

$$\text{T.M.} = \int \theta_i \theta_f \, d\tau_N . \int S_i S_f \, d\tau_s . \int \varphi_i . \boldsymbol{\mu} . \varphi_f \, d\tau_e \tag{2.11}$$

The first term is the overlap integral of the wavefunctions for nuclear vibrations—it embodies a quantum mechanical formulation of the Franck–Condon principle (see p. 17). The second term is a spin overlap integral, and its value depends on the initial and final spin states of the promoted electron. The third term is called the *electronic transition moment*, and its value depends on the symmetries and amount of overlap of the initial and final spatial orbitals.

The previous discussion shows that if the Born–Oppenheimer approximation is valid, the large and insoluble Schrödinger equation, $H\Psi = E\Psi$, can be split into three simpler equations:

$$H\varphi = E\varphi$$

$$H\theta = E\theta$$

$$HS = ES$$

and the total molecular energy can be expressed as the sum of electronic, vibrational and spin energies:

$$E = E_e + E_v + E_s$$

The Born–Oppenheimer approximation is a good one for most purposes, but it is important to realize (i) that it breaks down near degeneracies, i.e. where potential energy surfaces cross, for in this region a small change in nuclear coordinates can take a molecule from one surface to the other and thus produce a large change in the electronic wavefunction, and (ii) that it fails to account for the non-zero intensities observed with forbidden transitions (see Vibronic coupling, p. 25).

2.1.4 Selection Rules

The transition moment in the approximate form given in equation (2.11) is seen to be the product of three separate integrals, and its value will be zero if any one of the component integrals is zero. When this happens, the transition has a zero probability of occurrence. Such a transition is said to be *forbidden*, in contrast to *allowed* transitions in which the transition moment is non-zero. Forbidden transitions are only totally forbidden within the framework of approximations implicit in the derivation of equation (2.11). More refined calculations show that forbidden transitions should have small intensities. Thus, they can usually still be observed, although with intensities much less than those of allowed transitions.

Using equation (2.11), the transition moment can be evaluated and hence the approximate oscillator strength can be predicted. This, though not difficult with the aid of modern computer programs, is often unnecessary, because it is sufficient to know whether the transition is allowed or forbidden, and selection rules permit us to determine this point without detailed calculation.

Selection rules derive from symmetry considerations, which often indicate whether, for a particular transition, the integrand of one of the component integrals is an *odd* (antisymmetric) function of the coordinates. In such a case, the integral will be zero and the transition will be forbidden.

To illustrate this point, consider Figures 2.1 and 2.2 exhibiting typical even and odd functions. For an odd (antisymmetric) function, it is clear that for each individual positive infinitesimal area $f(x)\,dx$ there is a corresponding negative contribution. Thus the integral $\int_{-\infty}^{+\infty} f(x)\,dx$, which is merely the sum of these infinitesimal contributions to the area under the curve, will be zero. For an even (symmetric) function, the integral will be non-zero. With this in mind, the three component integrals of the transition moment integral will be examined.

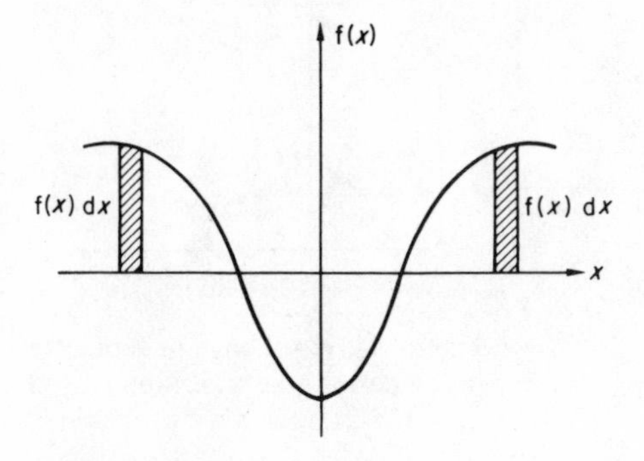

Figure 2.1. Symmetric (even) function of x

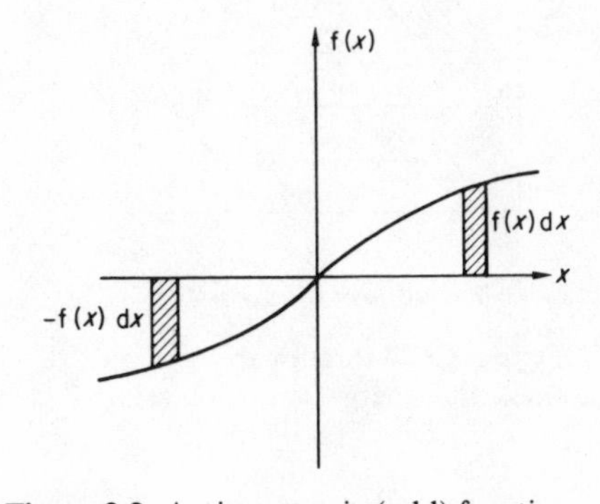

Figure 2.2. Antisymmetric (odd) function of x

2.1.5 Vibrational Overlap Integral (Franck–Condon Principle)

In order to evaluate the integral $\int \theta_i \theta_f \, d\tau_N$, it is necessary to know the vibrational wavefunctions. For simplicity, consider a simple harmonic oscillator. For such a system, the potential energy for varying internuclear separations is described by the parabolic curve of Figure 2.3, the equilibrium separation (r_e) being given by the point (a) where the potential energy is a minimum. Solution of the appropriate Schrödinger equation leads to the result that the vibrational energy is quantized with values

$$E = h\nu(v + \tfrac{1}{2})$$

where v, the vibrational quantum number, must be integral $0, 1, 2, \ldots, n$. These values are represented by the equally spaced horizontal lines in Figure 2.3. (Note that the minimum vibrational energy is not zero but $\frac{1}{2}h\nu$, and that at room temperature in condensed phases most molecules will be in the $v'' = 0$ level). The oscillating curves represent the wavefunctions associated with each vibrational level. Since the square of the amplitude (θ) of the wavefunction gives the probability of a particular nuclear configuration, we can see that in the bottom level ($v'' = 0$) the nuclei are most likely to be found at the equilibrium distance r_e, and that as v'' increases it becomes increasingly probable that the molecule will be found close to the configuration corresponding to the intersection of the horizontal lines and parabolic curve (the turning point of the vibration, where the total energy equals the potential energy, the kinetic energy is zero, and the nuclei are static).

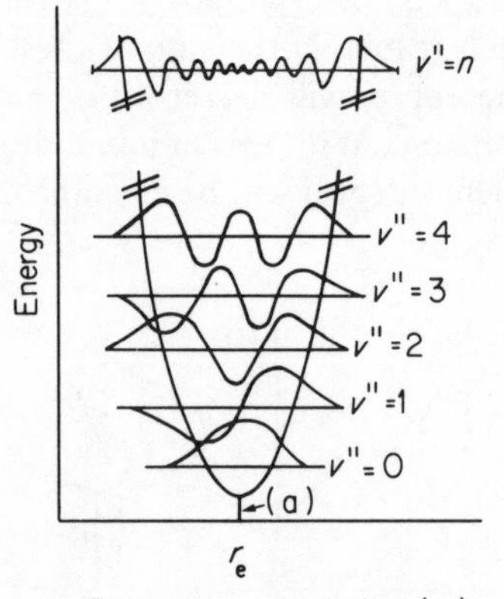

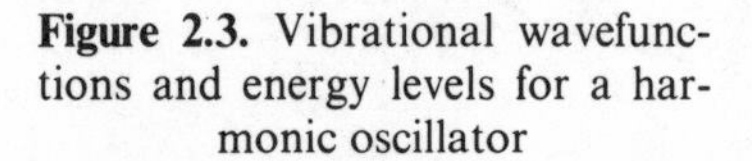
Figure 2.3. Vibrational wavefunctions and energy levels for a harmonic oscillator

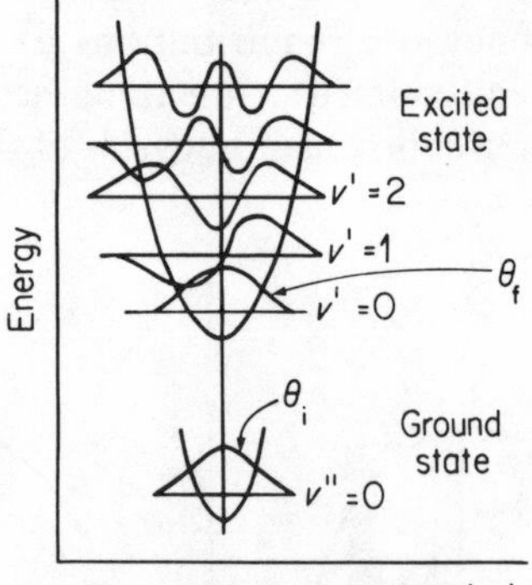

Figure 2.4. No change in geometry on excitation; the transition $v'' = 0$ to $v' = 0$ is the most probable

In Figure 2.4 are represented the potential energy curves for the ground and excited states of such an idealized molecule. In this case there is no change in geometry on excitation (r_e is the same for both curves). To evaluate the vibrational overlap integral, we multiply the individual values of the wavefunctions (θ_i and θ_f) for each internuclear separation and sum the infinitesimal contributions $\theta_i\theta_f \, . \, dr$. It is clear from the Figure that the value of the vibrational overlap integral is positive for the transition $v'' = 0 \rightarrow v' = 0$ [a so-called $(0 \rightarrow 0)$ transition], and zero for the $0 \rightarrow 1$ transition because for each positive infinitesimal contribution to the overlap integral there is an equal negative contribution. The same is true for transitions to upper odd vibrational levels, and almost true for transitions to even levels. Hence even though real molecules are not harmonic oscillators, when there is only a small change in geometry on excitation the $(0 \rightarrow 0)$ vibrational transition is expected to be the strongest; $(0 \rightarrow n)$ transitions will be much less probable and thus less intense.

Figure 2.5 represents the more common situation where excitation leads to stretching of a bond. This is to be expected wherever the excited electron is

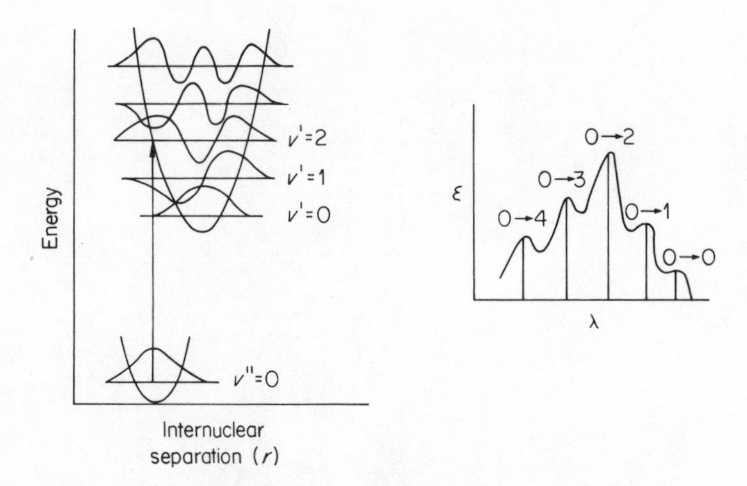

Figure 2.5. Change in geometry on excitation, leading to the vibrational fine structure in the spectrum shown

promoted into an antibonding orbital, thereby weakening the bond. In this particular example, inspection reveals that the $(0 \rightarrow 2)$ vibrational transition will be strongest.

Similar considerations apply to emission spectra, with the proviso that in condensed phases at room temperature the excited state will normally become thermally equilibrated before emission, which therefore takes place almost exclusively from the bottom vibrational level of the excited state.

The above discussion, based on the vibrational overlap integral, is a quantum mechanical formulation of the *Franck–Condon Principle* first expressed classically by Franck. The principle states that electronic transitions occur in an exceedingly brief interval of time so that no change in nuclear position or nuclear kinetic energy occurs during the transition. This implies that the transition may be represented by a *vertical* line connecting the two potential energy surfaces, and the most probable transition will be to that vibrational level with the same internuclear distance at the turning point of the oscillation, e.g. lines AY and ZB in Figure 2.6. A transition represented by line AX is extremely improbable because the molecule, in arriving at point X, would have suddenly acquired an excess kinetic energy given by XY.

It must be emphasized that real molecules are *anharmonic* oscillators, and for diatomics the potential energy diagram is approximated by a Morse curve (Figure 2.7), the higher vibrational levels of which become progressively closer together until they merge into a continuum at the dissociation limit. Furthermore, for any molecule containing more than two atoms the potential energy curve becomes a surface in many dimensions because of the very large number of possible vibrational modes. Hence corresponding to a particular electronic transition there will be a multiplicity of associated vibrational and rotational transitions so closely spaced that they overlap, giving rise to a smooth broad

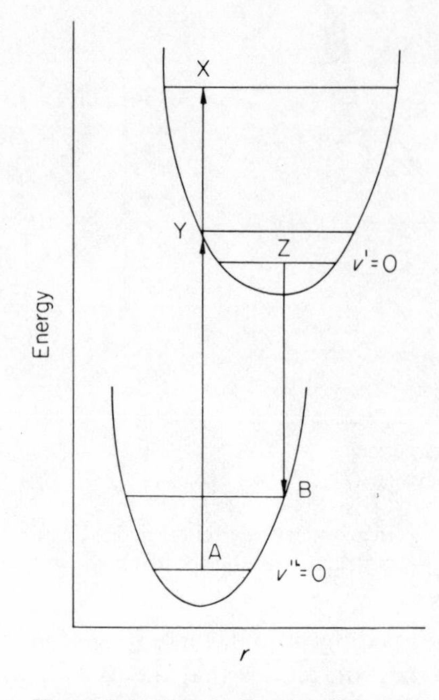

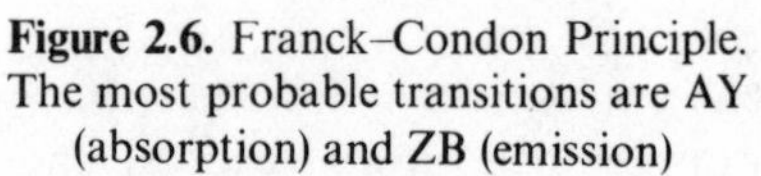
Figure 2.6. Franck–Condon Principle. The most probable transitions are AY (absorption) and ZB (emission)

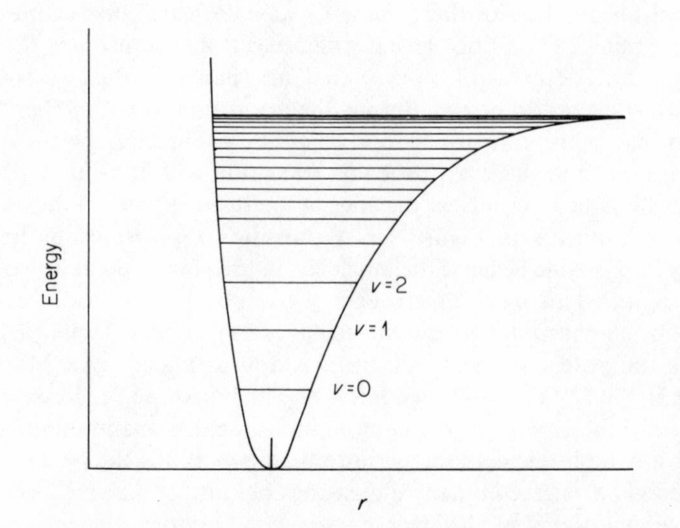

Figure 2.7. Energy diagram for a diatomic molecule (anharmonic oscillator)

absorption. Therefore in the majority of cases little vibrational structure may be seen in a visible or ultraviolet absorption band. Even so, the shape of the absorption band is determined by the Franck–Condon principle, and its envelope is for this reason known as the Franck–Condon envelope.

In those cases where vibrational structure may be discerned,† important information about the shape of an excited molecule and its vibrational modes may be gleaned from an analysis of this *vibronic* (*vibr*ation + electr*onic*) structure.

2.1.6 Spin

The effect of electron spin upon transition intensities is given by the factor $\int S_i S_f \, d\tau_s$ in the transition moment expression. There are three common situations.

(a) *Singlet ⟶ singlet transitions.* The electron can have only two spin states (wavefunctions designated α and β). In a singlet ⟶ singlet transition no change occurs in the spin state of the promoted electron, and the spin overlap integral is $\int \alpha\alpha \, d\tau_s = \int \beta\beta \, d\tau_s = 1$ because the spin wavefunctions are assumed to be normalized. There are no spin restrictions on such transitions, which are therefore fully allowed.

(b) *Triplet ⟶ triplet transitions.* Since the transition occurs with no change in multiplicity, again the spin overlap integral $\int \alpha\alpha \, d\tau_s = 1$ and the transition is fully allowed. This fact is used in *flash photolysis* (see Chapter 5, p. 137), when triplet species are produced in high concentration by intersystem crossing after an intense burst of light and are examined by absorption spectroscopy.

(c) *Singlet ⟶ triplet transitions.* Since the promoted electron changes its spin state the spin overlap integral is $\int \alpha\beta \, d\tau_s = \int \beta\alpha \, d\tau_s = 0$, because the α and β spin wavefunctions are orthogonal. The transition moment is thus zero and the transition is strongly forbidden.

Spin-orbit Coupling

Singlet ⟶ triplet transitions are strongly forbidden; that they occur at all is due to spin–orbit coupling. Since the electron is charged and 'spinning', it is expected to have not only spin angular momentum but also a magnetic moment. In a $S \rightarrow T$ transition, an electron inverts its spin, i.e. changes the direction of its magnetic moment. Clearly this calls for a magnetic interaction (consider trying to change the orientation of a bar magnet without touching it). The required magnetic interaction is provided by the magnetic field produced by the *orbital* motion of the charged electron. The magnetic moment of the *spinning* electron becomes coupled to the orbital magnetic field—hence the term spin–orbit coupling.

The effect may be seen more clearly by basing the co-ordinate system on the electron. Seen from the electron, there is a charged nucleus in orbit around it

† Vibrational fine structure is best observed in the gas phase or in solution in non-polar solvents where perturbations in the energy of the solute caused by intermolecular interactions are minimized.

which is equivalent to a ring of current. This current generates a magnetic field that interacts with the magnetic moment of the electron, giving rise to spin–orbit coupling. The magnitude of the coupling constant increases very rapidly with increasing nuclear charge. This simple treatment implies that although the total angular momentum (spin and orbital) is conserved, neither orbital nor spin angular momentum are individually conserved—that because of the spin–orbit coupling, angular momentum is being continually switched between the spin and orbital modes. This means that one can no longer think of pure spin states (singlet or triplet), but intermediate situations must be contemplated in which a nominal singlet state has a certain degree of triplet character and vice-versa.

The spin–orbit interaction is treated quantum mechanically by introducing into the Hamiltonian operator a term H_{SO} for each electron of the form:

$$H_{SO} = k\zeta(\mathbf{L}.\mathbf{S})$$

where $\mathbf{L}$ is the orbital angular momentum operator, $\mathbf{S}$ is the spin angular momentum operator, and ζ is a factor depending on the nuclear field. Perturbation theory shows that if Ψ_S^0 and Ψ_T^0 are the wavefunctions of 'pure' singlet and triplet states respectively, then the triplet state produced under spin–orbit coupling can be written[3] in the form in equation (2.12), where S_k is the kth singlet state, and E_T and E_S are the energies of the triplet and the perturbing singlet states respectively.

$$\Psi_T = \Psi_T^0 + \sum_k \frac{\langle \Psi_{S_k}^0 | H_{SO} | \Psi_T^0 \rangle}{(E_T - E_{S_k})} \cdot \Psi_{S_k}^0 \tag{2.12}$$

A similar expression can be written for the singlet state. Thus the effect of spin–orbit coupling is to mix a small amount of singlet character into the triplet states and vice-versa, so that 'pure' singlet and triplet states no longer exist. Singlet → triplet transitions can then be thought of as occurring between the pure singlet and pure triplet components of each hybrid state. Figure 2.8 represents the situation in diagrammatic form.

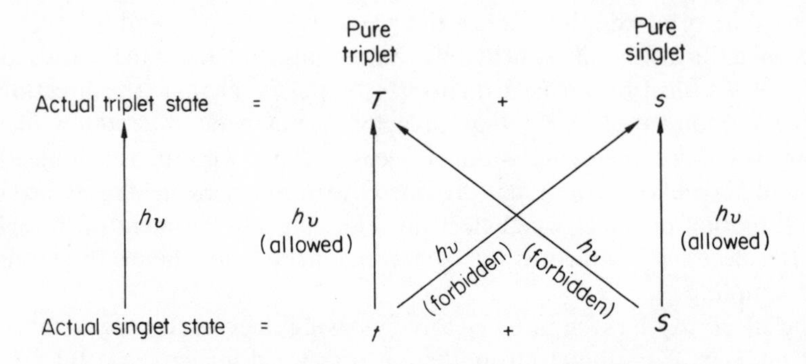

Figure 2.8. Diagrammatic resolution of actual states into their pure components. Forbidden $S \rightarrow T$ transitions appear as allowed transitions between the impurity components of the nominal states

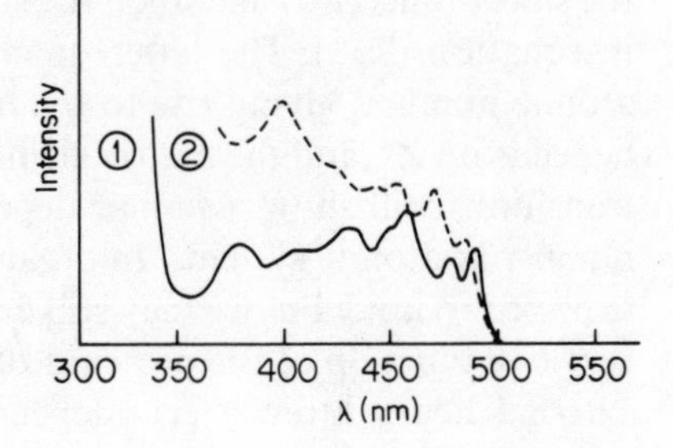

Figure 2.9. $S_0 \rightarrow T_1$ absorption spectra of (1) l-chloronaphthalene, and (2) l-iodonaphthalene (reproduced with permission from A. P. Marchetti and D. R. Kearns, *J. Amer. Chem. Soc.*, **89**, 768 (1967); copyright by the American Chemical Society)

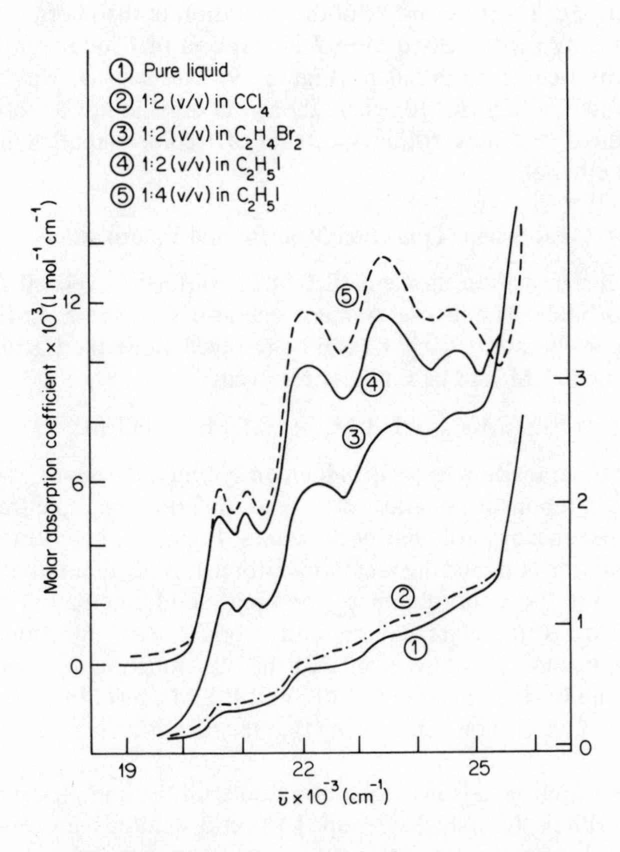

Figure 2.10. The effect of 'heavy-atom' solvents on the $S_0 \rightarrow T_1$ transition of 1-chloronaphthalene. Curves 1, 2 and 3 refer to the right-hand scale, curves 4 and 5 to the left-hand scale. (From S. P. McGlynn, T. Azumi and M. Kasha, *J. Chem. Phys.*, **40**, 507 (1964), reproduced by permission of the American Institute of Physics)

The probability of the $S \rightarrow T$ transition depends upon the energy gap between the states concerned and upon the size of matrix elements such as $\langle \Psi^0_{S_k} | H_{SO} | \Psi^0_T \rangle$ in equation (2.12). The latter quantity increases very rapidly with increasing atomic number, giving rise to the *heavy atom effect*. As explained earlier, H_{SO} depends on Z^4, and the matrix elements and therefore the probability of $S \rightarrow T$ transitions will show a similar dependence on the fourth power of the atomic number in atomic systems. In organic molecules this simple relation cannot be expected to hold, but we may still expect that $S \rightarrow T$ transitions will be a highly sensitive function of the presence of heavy atoms. There are both internal and external heavy atom effects depending upon whether the heavy atom is incorporated into the molecule itself or into the environment, but, in both cases, the presence of the heavy atom is manifested by an increase in the probability of singlet–triplet transitions, whether absorptive, emissive or radiationless (see Chapter 3, p. 82). An example of the operation of the internal heavy atom effect is given by the increased $S \rightarrow T$ absorption of 1-iodonaphthalene over that shown by 1-chloronaphthalene (Figure 2.9). The external heavy atom effect is demonstrated in Figure 2.10, where the use of ethyl iodide as solvent greatly intensifies the $S \rightarrow T$ absorption spectrum of 1-chloronaphthalene over that observed in ethanol.

2.1.7 Electronic Transition Moment and Polarization

The electronic transition moment (E.T.M.) is intimately related to the symmetries of orbitals. The dipole moment operator $\boldsymbol{\mu}$ in the E.T.M. $\int \varphi_i \boldsymbol{\mu} \varphi_f \, d\tau$, being a vector operator ($= e \sum \mathbf{r}_i$), can be resolved along the Cartesian axes of space, and the E.T.M. can be similarly resolved:

$$\text{E.T.M.}_{\text{total}} = \text{E.T.M.}_x + \text{E.T.M.}_y + \text{E.T.M.}_z$$

In order for the transition to be forbidden on symmetry grounds, it is necessary for all three component integrals to be zero. A simple example from the field of atomic spectroscopy will clarify the issues. Figure 2.11 illustrates the symmetry of s and p orbitals and the vector operator $\boldsymbol{\mu}$. It is clear that whereas ψ_{1s} is an even function of the x-coordinate, ψ_{p_x} and $\boldsymbol{\mu}$ are odd functions. The integrand of the E.T.M. is therefore even × odd × odd = even function,† and the E.T.M.$_x$ is non-zero. It is also clear that the wavefunction ψ_{p_x} has zero value along the y- and z-axes, making E.T.M.$_y$ and E.T.M.$_z$ zero. In such a situation, where only E.T.M.$_x$ is non-zero, we say that the transition is *polarized* along the x-axis.

This simple approach is inadequate to deal with the complex symmetries of molecular orbitals for which Group Theoretic methods are essential. The technique is described in the Appendix, but in essence it involves assigning symmetry symbols to ψ_i, ψ_f, $\boldsymbol{\mu}_x$, $\boldsymbol{\mu}_y$, and $\boldsymbol{\mu}_z$, and hence determining the symmetry of the integrands of the components of the E.T.M. Unless the integrand is

† The rule is odd × odd = even × even = even; odd × even = odd. Note that the terms symmetric and antisymmetric are frequently used instead of even and odd respectively.

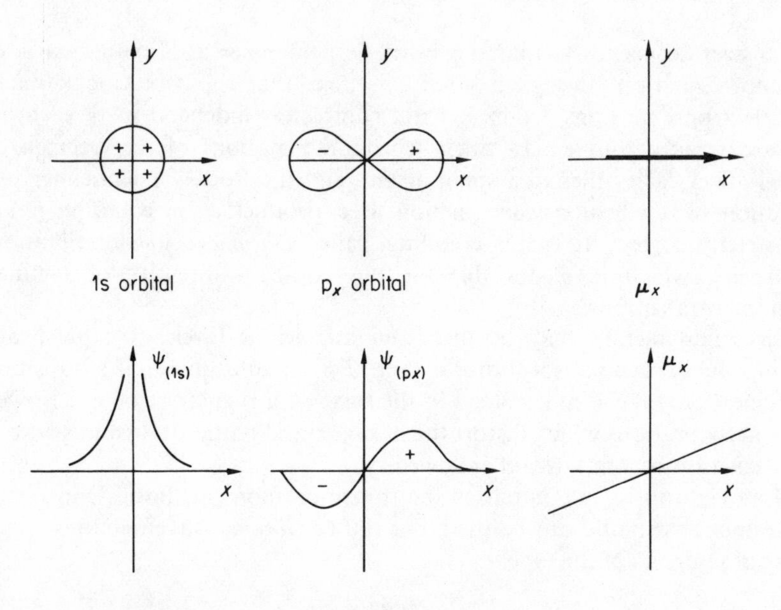

Figure 2.11. Symmetries of 1s and p_x orbitals and of $\boldsymbol{\mu}_x$

totally symmetric, the transition is forbidden. If the integrand of just one of the components of the E.T.M. (say E.T.M.$_x$) is totally symmetric, the transition will be polarized along the x-axis. For such a case, if we illuminate an oriented assembly of molecules (crystal or thin layer) with plane-polarized light (the electric vector of which oscillates in a particular plane), maximum absorption will occur when the plane of polarization of the incident light is parallel to the molecular x-axis. Rotation of the crystal will cause a decrease in absorption which falls to zero when the plane of polarization and the x-axis are perpendicular.

In a similar way, emission is frequently polarized along a particular molecular axis. Polarization measurements can give important information about the electronic transitions responsible for particular absorption bands.[4]

2.1.8 Vibronic Coupling and Forbidden Transitions

The previous discussion has shown that under the Born–Oppenheimer approximation the transition moment can be factorized (equation 2.11), so that if the electronic transition moment is zero, then the transition moment is zero and the transition will be forbidden. Therefore symmetry-forbidden transitions should have zero intensity and be unobservable; nonetheless, they are found to have small but finite intensities. Examples of such forbidden but observed transitions are the benzene absorption near 254 nm ($^1A_{1g} \rightarrow B_{2u}$), the $n \rightarrow \pi^*$ transitions of aliphatic aldehydes and ketones, and the $d \rightarrow d$ transitions of the metal atom in centrosymmetric complexes. The question then arises and to why such symmetry-forbidden transitions should be observable, and the answer seems to be that the Born–Oppenheimer approximation is only an approximation (though a good one); i.e. that the vibrational (nuclear) and electronic motions are in fact not completely independent of each other but are weakly coupled. In other words, the motions of the electrons and nuclei affect each other to a small extent, and the Born–Oppenheimer representation of a vibronic wavefunction as a product as in equation (2.11) is not strictly correct. It is this coupling, called *vibronic coupling* (*vibr*ational–electr*onic*), which is responsible for the non-zero intensity of symmetry-forbidden transitions.

This phenomenon may be discussed at various levels of sophistication. Taking the benzene absorption $^1A_{1g} \rightarrow {}^1B_{2u}$ as an example, the transition is forbidden only if the molecule is in the form of a regular hexagon. However, there are vibrations which distort the hexagon and reduce its symmetry so that the transition becomes weakly allowed.

More rigorously, in computing the transition moment, Born–Oppenheimer wavefunctions should not be used, but rather *vibronic* wavefunctions, and the value of integrals of the type

$$\langle \psi_i | \boldsymbol{\mu} | \Psi_f \rangle$$

should be assessed, where the Ψ terms are now vibronic wavefunctions. The value of the transition moment so obtained will not differ greatly from that estimated using Born–Oppenheimer wavefunctions, but for vibrations of the appropriate symmetries it will assume a small value when the Born–Oppenheimer transition moment is zero. Thus symmetry-forbidden transitions do occur, though with small intensities.

The symmetries of vibronic wavefunctions are dependent upon the symmetries of the component vibrations. This makes it possible, with the aid of Group Theory, to predict which vibrations are effective in making the transition 'slightly allowed' (see Appendix for a worked example). Vibronic coupling is also of importance in non-radiative transitions (see Chapter 3, p. 89).

2.1.9 Orbital Overlap

Suppose that the initial and final orbitals occupy such different regions of space that they have but little overlap. The overlap integral $\int \psi_i \psi_f \, d\tau$ is a measure of the amount of this spatial overlap, and it is evaluated by summing the infinitesimal contributions $\psi_i \psi_f \, d\tau$. The E.T.M. is obtained by multiplying each of these infinitesimals by the appropriate value of $\boldsymbol{\mu}$ before summing. Thus, if ψ_i and ψ_f are so oriented that large values of the wavefunction ψ_i occur where ψ_f is small and *vice versa*, then the product $\psi_i \psi_f$ will always be small, as will the triple product $\psi_i \boldsymbol{\mu} \psi_f$ and also the E.T.M. In such a situation, there will be a symmetry-allowed transition which is weak. An example is afforded by the $(n \rightarrow \pi^*)$ transition of pyridine. Group Theory (see Appendix) shows that this transition is allowed, but inspection of Figure 2.12 shows that the n and π orbitals occupy quite different regions of space. The transition is thus forbidden on overlap grounds, and it is not surprising that the extinction coefficient of the band is only 400 l mol^{-1} cm^{-1}. The $(n \rightarrow \pi^*)$ transition of formaldehyde ($\varepsilon_{max} \sim 20$ l mol^{-1} cm^{-1}) is forbidden on both symmetry and overlap grounds.

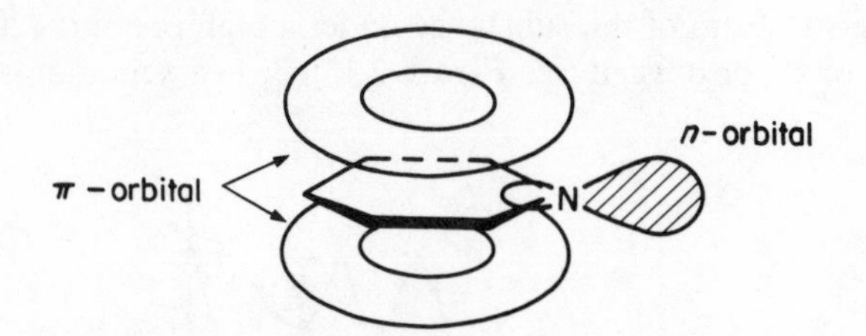

Figure 2.12. Limited overlap of π- and n-orbitals in pyridine

2.1.10 Oscillator Strengths and Forbidden Transitions

From the previous sections it is seen that a transition contravening a selection rule is forbidden, and the corresponding absorption band will be less intense than it would otherwise have been. The effect of such contraventions of the selection rules may be estimated very approximately as follows. If a completely allowed transition has an oscillator strength $F_A \sim 1$, then other transitions have oscillator strengths F given roughly by:

$$F = f_s \cdot f_o \cdot f_{sym} \cdot F_A$$

where the f factors correct for the varying degrees of forbiddenness:

f_s(spin) $\sim 10^{-8}$ for aromatic hydrocarbons, $\sim 10^{-5}$ for second-row elements,
f_o(orbital overlap) $\sim 10^{-2}$ for $(n \rightarrow \pi^*)$ transitions of first-row elements,
f_{sym}(symmetry) $\sim 10^{-1}$–10^{-3}.

2.1.11 Other Factors Affecting the Intensities of Absorption Spectra

Hot Bands

When vibrational fine structure can be discerned, the lowest energy (longest wavelength) vibrational band is normally due to the $(0 \rightarrow 0)$ transition, because it is the $v'' = 0$ level of the ground state which is predominantly populated at room temperature. At higher temperature, or when the $(0 \rightarrow 0)$ transition is weak (symmetry forbidden) as with benzene, weak absorption bands can be discerned at longer wavelengths than the $(0 \rightarrow 0)$ transition. These are ascribed to transitions from levels higher than the lowest vibrational level of the ground state, because their intensity increases with temperature as the population of these higher levels increases. Such bands are known as *hot bands*.

External Perturbations[5]

Heavy atoms incorporated either into the substrate molecule or into the solvent enhance $S \rightarrow T$ absorption through a spin–orbit coupling effect. Similarly, an increase in the intensity of $S \rightarrow T$ transitions may be induced by observing the spectrum of the substance under a high pressure (20–100 atm) of xenon, nitric oxide or oxygen (see Figure 2.13), or in a xenon or oxygen matrix

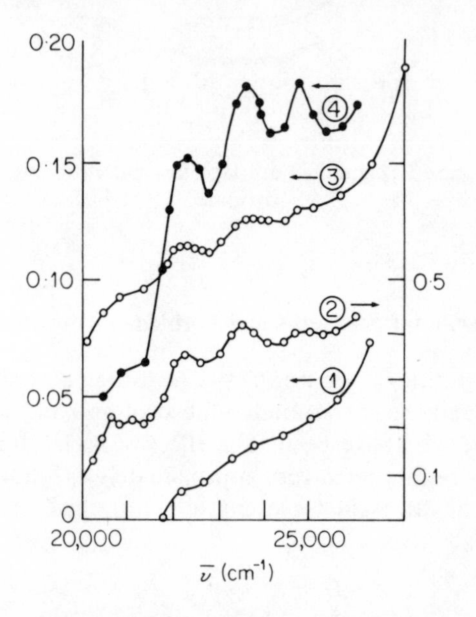

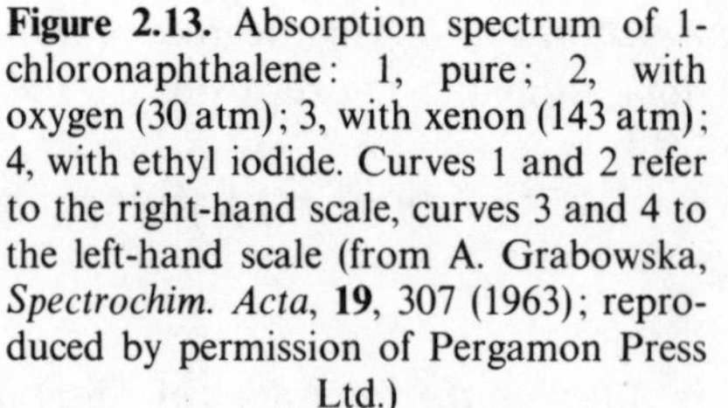

Figure 2.13. Absorption spectrum of 1-chloronaphthalene: 1, pure; 2, with oxygen (30 atm); 3, with xenon (143 atm); 4, with ethyl iodide. Curves 1 and 2 refer to the right-hand scale, curves 3 and 4 to the left-hand scale (from A. Grabowska, *Spectrochim. Acta*, **19**, 307 (1963); reproduced by permission of Pergamon Press Ltd.)

at 10 K. The xenon probably operates through the external heavy atom effect, but a different mechanism must be responsible for the enhanced absorption observed with oxygen and nitric oxide. Evans,[6] who devised the oxygen perturbation technique, ascribed it to the paramagnetism of the oxygen molecule. It now seems,[7] however, that paramagnetic interactions are probably negligible and that the absorption enhancement is to be attributed to a mixing of states within contact charge–transfer complexes (see p. 35) involving oxygen or nitric oxide. These molecules also enhance non-radiative $S \rightarrow T$ transitions (intersystem crossing, see Chapter 4, p. 116).

2.2 TYPES OF TRANSITIONS

The common types of molecular orbitals encountered in organic molecules were briefly described in Chapter 1. Electronic transitions between these orbitals give rise to different types of excited state, the nature of which can often be specified from an analysis of the absorption spectrum. Whether or not this initially produced excited species is the one responsible for a particular photochemical reaction depends on the magnitude of the rate constant for the chemical transformation relative to those for competing radiationless processes (internal conversion, intersystem crossing) which lead to alternative excited states. How this point is elucidated is discussed in Chapter 5. In this section the utilization of absorption spectra to identify the initial state is considered.

2.2.1 Nomenclature

Different systems of describing excited states are employed depending on which properties of the state are to be emphasized and on the amount of information available concerning the nature of the originating transition. Therefore, if symmetry considerations are dominant, group theoretical symbols will be adopted. For example, the ground state of benzene ($^1A_{1g}$) is excited by the 253·7 nm mercury line into its first electronically excited state, designated $^1B_{2u}$, and the transition would be symbolized $^1A_{1g} \longrightarrow {}^1B_{2u}$.† Alternatively, the molecular orbitals involved in the transition may be specified. In the benzene example, an electron is promoted from a π to a π^* orbital; the transition is then ($\pi \longrightarrow \pi^*$) and the state is symbolized (π, π^*). Since there are several π and π^* orbitals, clearly this nomenclature is less precise than that based upon group theory. In a similar manner, the long wavelength transition in aliphatic aldehydes and ketones can be designated ($n \longrightarrow \pi^*$) and it gives rise to an (n, π^*) excited state. This scheme is due to Kasha. When all that is at issue is the ordering of the states, a simple enumerative scheme is adopted. The singlet states are characterized $S_0, S_1 \ldots S_n$, where S_0 is the ground state and S_1 etc. are the higher singlets. The corresponding triplets are then $T_1, T_2, \ldots, T_n$. (There is no T_0 state except for some paramagnetic molecules such as oxygen.) Other schemes are occasionally found in the literature, notably that of Platt[8] for the characterization of the states of aromatic molecules, but they will not be used in this book.

The commoner transitions encountered in organic molecules are: $\pi \longrightarrow \pi^*$, $n \longrightarrow \pi^*$, $n \longrightarrow \sigma^*$, charge–transfer and Rydberg.

2.2.2 $\pi \longrightarrow \pi^*$ Transitions

(a) Ethylene provides the simplest example of a ($\pi \rightarrow \pi^*$) transition. The energetic ordering of the levels (Figure 2.14) makes this the lowest energy (longest wavelength) transition.

† It is a convention among spectroscopists to write the upper state first regardless of the direction of the arrow, so that $^1B_{2u} \longleftarrow {}^1A_{1g}$ represents absorption of radiation but $^1B_{2u} \longrightarrow {}^1A_{1g}$ designates fluorescence. Photochemists, however, tend to write the initial state or orbital first, irrespective of the ordering of the energy levels.

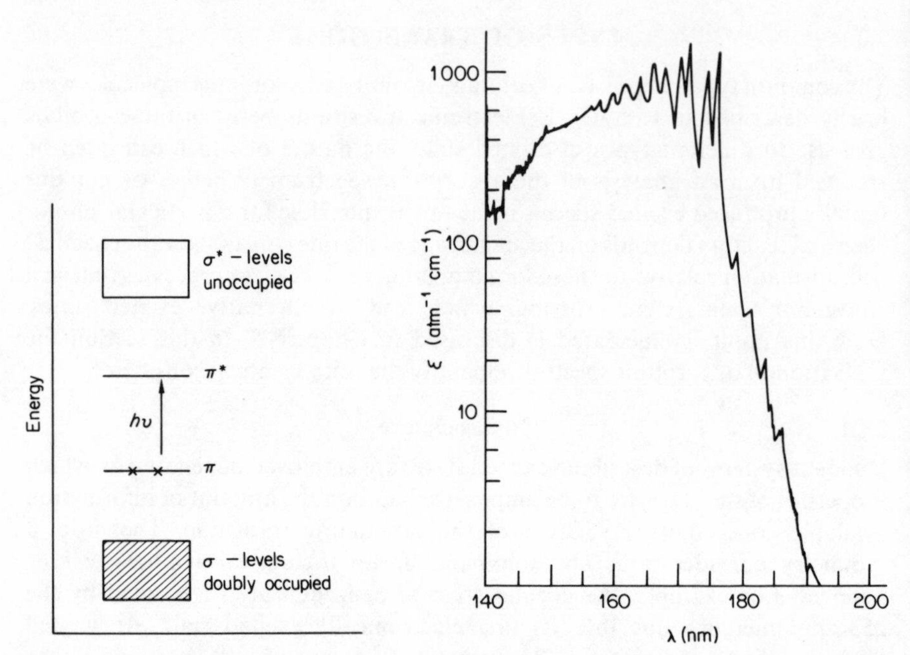

Figure 2.14. Energy levels of ethylene

Figure 2.15. Ultraviolet absorption spectrum of ethylene (from M. Zelikoff and K. Watanabe, *J. Opt. Soc. Amer.*, **43**, 756 (1953), reproduced by permission of the Optical Society of America)

The transition is allowed and gives rise to an intense absorption ($f \sim 0{\cdot}3$) in the vacuum ultraviolet. The absorption is broad and extends from ~137·5 to ~200 nm with superimposed sharp lines (Figure 2.15) which seem[9] to be the first members of a *Rydberg* series and in which a π electron is promoted into an orbital probably of σ-type and embracing a central $C_2H_4^+$ ion.

The triplet ($\pi \longrightarrow \pi^*$) transition is so weak ($\varepsilon \sim 10^{-4}\,\mathrm{l\,mol^{-1}\,cm^{-1}}$) that a path length of 14 metres of liquid ethylene is required for its detection. It gives rise to a band with $\lambda_{\max}$ at 270 nm and extending as far as 325 nm.

(b) Conjugation shifts the ($\pi \longrightarrow \pi^*$) transition to longer wavelengths (Figure 2.16). In molecular orbital terms, the phenomenon arises because the energy gap between the highest occupied molecular orbital (HOMO) and the lowest unoccupied molecular orbital (LUMO) becomes progressively smaller as the number of conjugated double bonds increases. Within the Hückel approximation the energy of the rth one-electron orbital adopts[10] the analytical form

$$E_r = \alpha + 2\beta \cos \frac{r\pi}{2n+1} \tag{2.13}$$

where n is the number of conjugated double bonds, and α and β are the coulomb and resonance integrals for carbon. Hence the energy of the longest wavelength transition absorption band is

$$\Delta E = E_{n+1} - E_n = -4\beta \sin \frac{\pi}{4n + 2} \tag{2.14}$$

This is a quantity which diminishes as n increases. Figure 2.17 shows how the interaction of two ethylenic double bonds in butadiene leads to a *bathochromic* (longer wavelength) displacement of the low energy transition of butadiene relative to that of ethylene.

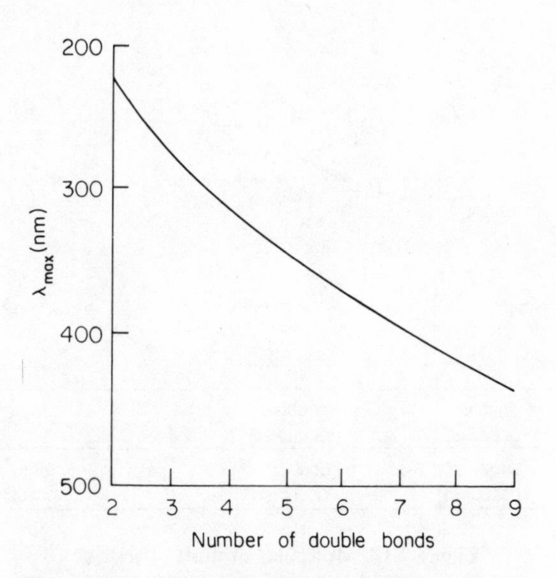

Figure 2.16. Ultraviolet/visible absorption maximum for the lowest energy transition in α,ω-dimethylpolyenes

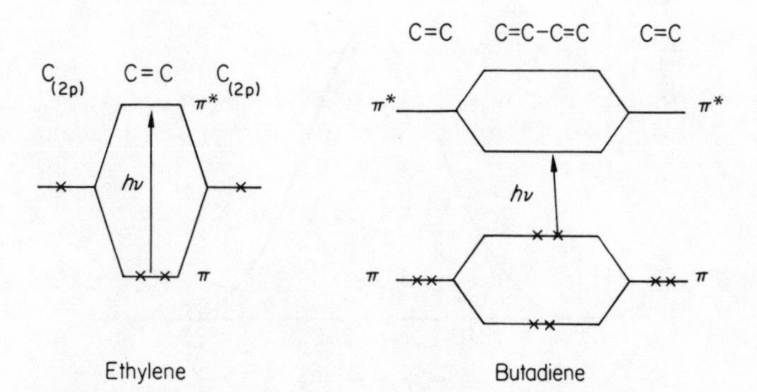

Figure 2.17. Orbital energies for ethylene and buta-1,3-diene

2.2.3 $n \rightarrow \pi^*$ Transitions

($n \rightarrow \pi^*$) Absorption is characteristic of molecules possessing chromophores with multiply bonded hetero-atoms (e.g. C=O, C=N, C=S, N=N, N=O). It is of low intensity ($f \sim 10^{-2}$–10^{-4}, $\varepsilon \sim 10$–$100\,\mathrm{l\,mol^{-1}\,cm^{-1}}$) being symmetry and/or overlap forbidden, and it is normally the band occurring at longest wavelength. Acetone will serve as an example. The relevant orbitals are shown in Figure 2.18 and the ultraviolet spectrum in Figure 2.19. The ($n \rightarrow \pi^*$)

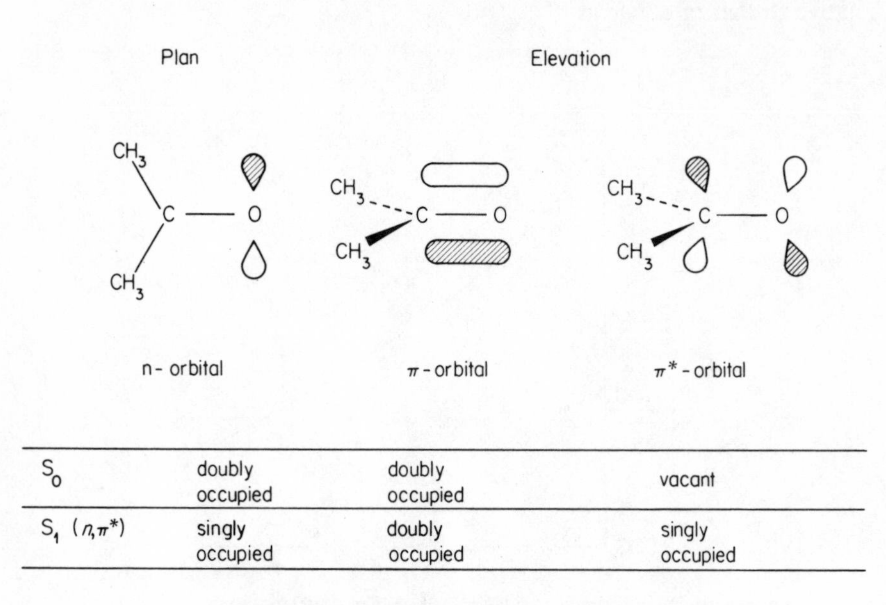

S_0	doubly occupied	doubly occupied	vacant
S_1 (n,π^*)	singly occupied	doubly occupied	singly occupied

Figure 2.18. Molecular orbitals of acetone

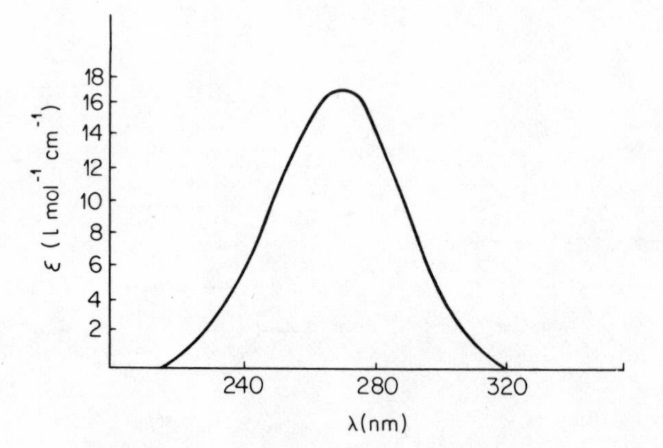

Figure 2.19. Absorption spectrum of acetone

transition has λ_{max} at ~ 280 nm. In the region of 190 nm is an intense band which may be due to $(n \rightarrow \sigma^*)$[11] or $(\sigma \rightarrow \pi^*)$ transitions,[12] and on the long wavelength tail of the $(n \rightarrow \pi^*)$ absorption are several weak bands due to the triplet $(n \rightarrow \pi^*)$ transition.

2.2.4 $n \rightarrow \sigma^*$ Transitions

The first absorption band of alkyl halides is due to the promotion of a non-bonding halogen p-electron into the σ^* antibonding C—X orbital. The transition is partially forbidden, and the absorption is weak ($\varepsilon_{max} \sim 200$ l mol^{-1} cm^{-1} for CH_3Cl). The promotion of an electron into the σ^* anti-bonding orbital largely neutralizes the bonding induced by the two electrons in the carbon–halogen σ-orbital, so that efficient dissociation is a consequence of irradiating alkyl halides in the near ultraviolet ($RX \rightarrow R^{\cdot} + X^{\cdot}$).

Other compounds with hetero-atoms singly linked to carbon (R—OH, R—SH, R—NH_2 etc.) also show $(n \rightarrow \sigma^*)$ absorption, but this tends to occur at wavelengths shorter than 200 nm (e.g. for MeOH, $\lambda_{max} = 183$ nm, $\varepsilon \sim 500$ l mol^{-1} cm^{-1}). As might be expected, there is a correlation between the ionization potential of the compound (involving ionization of the n-electron) and λ_{max}—the lower the ionization potential, the greater λ_{max}. For saturated compounds, as well as for alkenes, the importance of Rydberg states, in which an electron is promoted to an orbital of higher principal quantum number, is being increasingly recognized.[13]

2.2.5 Charge–Transfer (CT) Transitions

Mixtures of electron donors and electron acceptors in solution often exhibit a new absorption band which is shown by neither component separately and which is attributed to the presence in such mixtures of a donor–acceptor complex (DAC). Typical acceptors are picric acid and other polynitro aromatics, maleic anhydride, quinones and iodine, and typical donors include aromatic hydrocarbons and their derivatives, dienes and amines. The transition is referred to as a charge–transfer (CT) transition and in general is broad and structureless (see Figure 2.20). The nature of the bonding in DACs and the theory of CT absorption has been worked out by Mulliken.[14] Briefly, the ground state of a DAC is described by a wavefunction:

$$\psi_{(S_0)} = a\psi_{(DA)} + b\psi_{(D^+A^-)} \quad (2.15)$$

in which $\psi_{(DA)}$ corresponds to a 'no-bond' structure, the components being held only by weak intermolecular forces such as hydrogen bonding and dipole–dipole interaction, and $\psi_{(D^+A^-)}$ corresponds to the structure in which an electron has been totally transferred from the donor to the acceptor. The wavefunction for the excited states of the DAC is then given by:

$$\psi_{(S_1)} = a\psi_{(D^+A^-)} - b\psi_{(DA)} \quad (2.16)$$

For the majority of DACs $b \ll a$, so that $\psi_{(S_0)} \sim \psi_{(DA)}$, and $\psi_{(S_1)} \sim \psi_{(D^+A^-)}$.

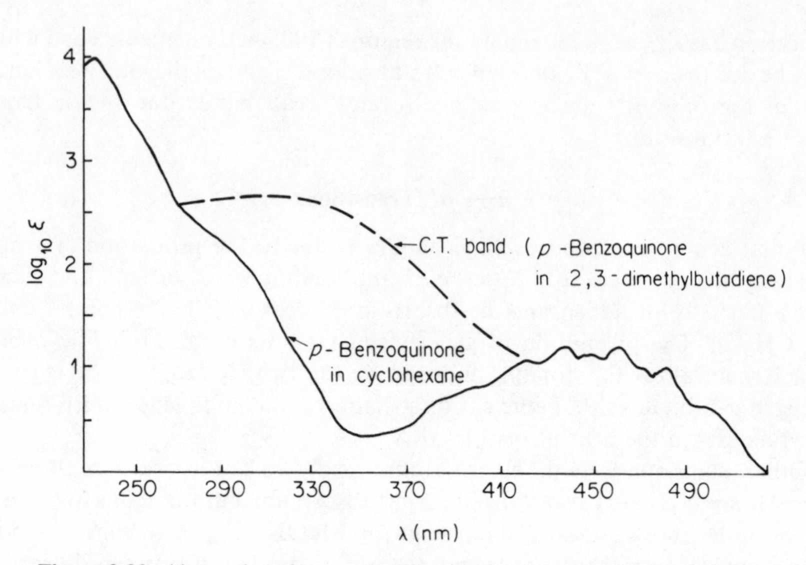

Figure 2.20. Absorption due to charge–transfer complex *p*-benzoquinone–2,3-dimethylbutadiene

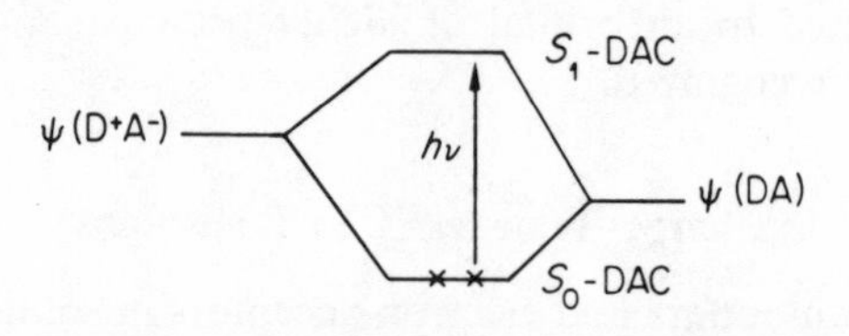

Figure 2.21. Stabilization of the ground state and destabilization of the excited state of a DAC, induced by mixing of states

Thus the spectroscopic transition corresponds approximately to the light-induced transfer of an electron from the donor to the acceptor, and hence the name charge–transfer transition. It should be noted that in such cases there is very little charge–transfer in the ground state, and it is perhaps inappropriate to call them charge–transfer complexes.

The energetics of DACs can be explored with the aid of energy diagrams. Figure 2.21 shows how mixing the wavefunctions of $\psi_{(DA)}$ and $\psi_{(D^+A^-)}$ leads to stabilization of the ground state and destabilization of the excited state. In Figure 2.22 ΔH_0 and ΔH_1 are the energies of formation of the ground state of a DAC from D and A, and of the excited state from D^+ and A^-, A_A and I_D are the electron affinity of the acceptor and the ionization potential of the donor. It is then clear that the energy of the CT transition (ΔE_{CT}) is given by

$$\Delta E_{CT} = I_D - A_A - (\Delta H_1 - \Delta H_0)$$

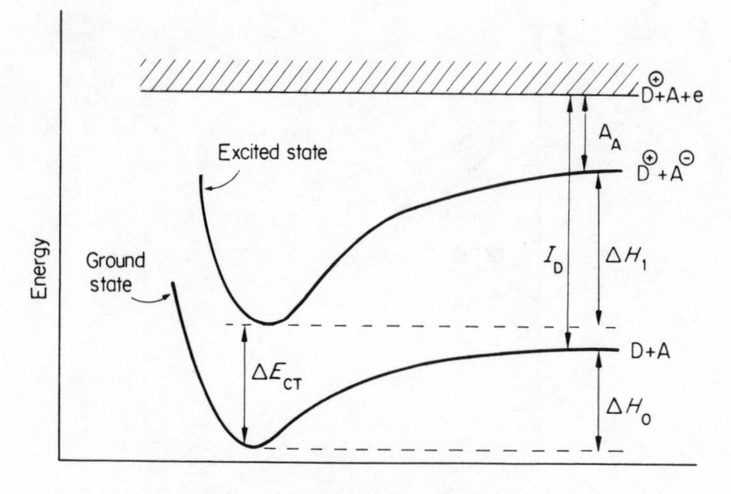

Figure 2.22. Energetics of donor–acceptor complexes

Thus, λ_{max} of the CT transition moves to longer wavelength as the donor and acceptor powers of the components increase. Extensive investigations[15] confirm the general validity of these principles.

2.2.6 Contact DACs and Contact Charge-Transfer Absorption

The phenomenon of contact CT absorption became known through the observations of Evans[16] that iodine in saturated hydrocarbons shows a new structureless absorption in a region where both components are transparent (Figure 2.23). The optical density of the band is proportional to both iodine and hydrocarbon concentrations, showing that the concentration of the DAC is extremely small. This point is borne out by a Benesi–Hildebrand[17] analysis which shows that the equilibrium constant for complex formation is zero. The data are consistent with the hypothesis that the transition occurs during a collision between donor and acceptor.

A number of other systems exhibiting contact charge–transfer absorptions are now known, notably oxygen in solution in saturated and aromatic hydrocarbons and in alcohols, ethers and even water. The basic difference between ordinary DACs and contact DACs seems to be twofold:

(i) the binding energy of contact DACs is very much smaller ($<kT$, where k is Boltzmann's constant), and

(ii) that, whereas ordinary DACs have definite structures, in contact DACs the two components are oriented randomly.

The theory of the spectra of contact CT absorption has been worked out by Murrell.[18]

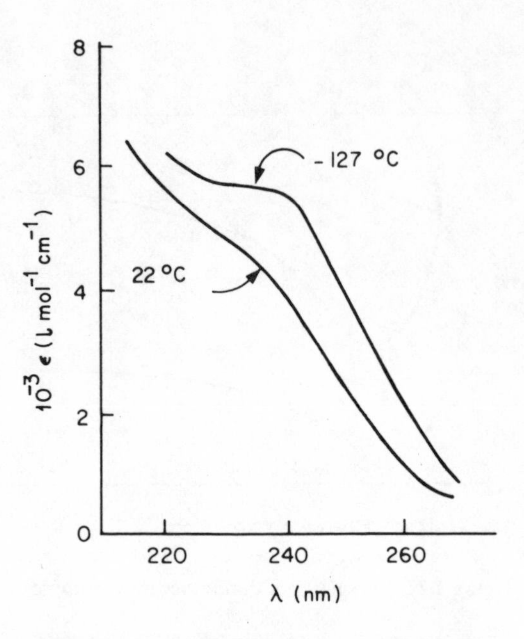

Figure 2.23. Absorption spectrum of iodine in methylcyclohexane. Contact charge–transfer band appears at ~240 nm in low temperature spectrum (from D. F. Evans, *J. Chem. Soc.*, 4229 (1957); reproduced by permission of the Chemical Society)

2.3 OTHER METHODS OF PRODUCING EXCITED STATES

2.3.1 Electrical Discharges

If a high potential is applied across an inert gas containing mercury vapour, the few stray electrons present are accelerated to energies sufficient to ionize the gas. Recombination of ions and electrons then gives rise to excited atoms of the inert gas. Direct excitation of the gas can also occur.

$$\mathrm{Ne} + e^- \rightarrow \mathrm{Ne}^+ + 2e^-$$

$$\mathrm{Ne}^+ + e^- \rightarrow \mathrm{Ne}^*$$

The excited atoms on collision with Hg atoms excite the latter (a phenomenon known as *energy transfer*, see Chapter 4), and these subsequently emit radiation:

$$\mathrm{Ne}^* + \mathrm{Hg} \rightarrow \mathrm{Ne} + \mathrm{Hg}^*$$

$$\mathrm{Hg}^* \rightarrow \mathrm{Hg} + h\nu$$

This is the basis for the functioning of mercury arc lamps so extensively used in photochemistry.

The spectral distribution of the emission of mercury arcs is markedly dependent on the pressure.

(a) Low pressure arcs, which contain Hg at pressures $\sim 10^{-3}$ mm, emit predominantly (95%) in the resonance line at 253·7 nm [$Hg(^3P_1) \rightarrow Hg(^1S_0)$] and to a much lesser extent at 184·9 nm [$Hg(^1P_1) \rightarrow Hg(^1S_0)$], see Figure 2.24.

(b) Medium pressure arcs are operated at Hg pressures in the region of 1 atmosphere with much higher currents, and this has the effect of raising the concentration of electrons, ions and excited species so that the metastable (relatively long-lived) $Hg(^3P_1)$ atoms, which would otherwise radiate at 253·7 nm, are promoted to many higher levels. Emission from these other states then gives rise to the multiplicity of lines characteristic of this type of arc. This rich and intense spectral output makes the medium pressure Hg arc an extremely important light source for photochemical investigations.† For physical measurements, where spectral purity is important, monochromators would be used to isolate particular mercury lines, but for preparative work, where it is often sufficient to use light embracing a range of wavelengths, filters are more appropriate. Compilations of filters are to be found in books by Calvert and Pitts[13] and by Bowen.[19]

(c) High pressure arcs, which run at pressures of up to several hundred atmospheres, are constructed with the electrodes close together thereby restricting the discharge to a small volume, and they constitute near point sources of brilliance comparable with that of the sun. The emission from such arcs is almost continuous. The continuum arises partly from pressure and temperature (Doppler effect) broadening of the numerous lines which consequently tend to overlap. In addition, excited Hg* atoms form excited dimers (excimers, see Chapter 4, p. 103) with Hg atoms in their ground state:

$$Hg^* + Hg \longrightarrow Hg_2^* \longrightarrow Hg + Hg + h\nu$$

Since the excimers are dissociated in their ground state, which therefore has no quantized levels, the potential diagram resembles that of Figure 2.25, and emission will clearly be continuous.

Other metals, such as Na, Zn, or Cd, which have appreciable vapour pressure at relatively low temperatures, may be used in place of Hg in discharge lamps with concomitant changes in the spectral output. In particular, Zn and Cd arc lamps have strong emission in the region 200–230 nm‡ in which the Hg arcs are notably deficient. High pressure arcs in inert gases, particularly xenon, are important sources of continuous radiation from <200 nm to the infrared. It should be noted that whereas electrical discharges provide an important way of exciting stable entities such as atoms, they are useless for systems of greater

† The long lifetime of $Hg(^3P_1)$ atoms means that the 253·7 nm line is very narrow both in emission and in absorption (Uncertainty Principle). However, in medium or high pressure Hg arcs the emission is broadened by pressure and temperature (Doppler effect), and the centre of the line at 253·7 nm is virtually absent (*line reversal*) because light generated in the body of the discharge is absorbed by 'cool' Hg atoms near the walls. This means that light from medium or high pressure arcs is useless for exciting Hg atoms, as in Hg sensitization experiments (see Chapter 8, p. 241), for which low pressure Hg arcs must be used.

‡ Strong ultraviolet emission from Zn vapour occurs at 214, 307 and 330–334 nm, and from Cd at 229 and 326 nm.

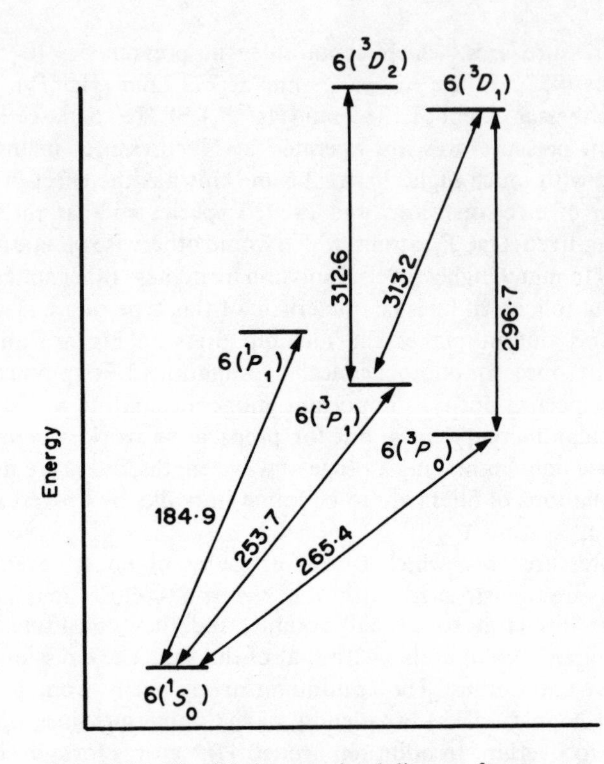

Figure 2.24. Simplified energy level diagram for mercury, showing a few of the states and the genesis of some of the more important mercury lines (in nm)

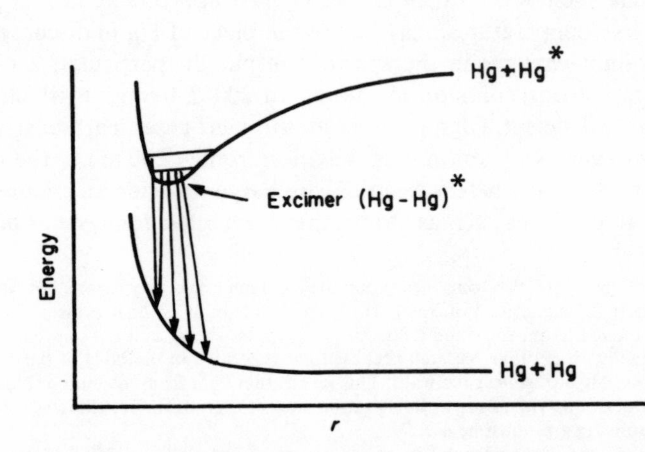

Figure 2.25. Continuous emission from mercury excimer

complexity. Diatomics are dissociated, and organic molecules are rapidly destroyed.

2.3.2 Ionizing Radiation

Media exposed to α, β or γ radiation become ionized, and the ensuing recombination reactions generate excited states, often in high yield. The phenomenon is exploited in pulse radiolysis (the ionizing radiation equivalent of flash photolysis) and in the scintillation detection and counting of such radiation. Here, an ionizing particle excites molecules of an organic solvent, energy transfer (see Chapter 4) to fluorescent solute molecules excites the latter, and the resulting emission is detected as a pulse by a photomultiplier.

2.3.3 Thermal Activation

Simple consideration of the Boltzmann distributed law:

$$\frac{N_u}{N_1} = e^{-\Delta E/RT}$$

where N_u and N_1 are the number of molecules in the upper and lower states and ΔE is their energy separation, reveals the virtual impossibility of producing electronically excited species by thermal methods except at extreme temperatures. If $\Delta E = 209$ kJ mol^{-1} (50 kcal mol^{-1}), then at 20 °C

$$\frac{N_u}{N_1} = e^{-209\,000/(8{\cdot}31 \times 293)} = 2{\cdot}4 \times 10^{-38}$$

but at 5000 °C

$$\frac{N_u}{N_1} = 6{\cdot}2 \times 10^{-3}.$$

Such temperatures may be achieved in flames, and even higher temperatures may be generated in shock tubes, but since these temperatures destroy organic molecules, thermal excitation is largely confined to atoms and diatomics.

2.3.4 Chemical Activation (Chemiluminescence)[20]

The excess energy of product molecules formed in chemical reactions is normally dissipated by collisions with the environment. Occasionally, however, in strongly exothermic reactions, the energy of the reaction is trapped as electronic energy, so that one of the products is formed in an electronically excited state. If this product is fluorescent, then radiation is emitted and luminescence is observed. Such chemically induced emission is known as *chemiluminescence*. The more efficient chemiluminscent systems fall into three groups:

(a) *Electron Transfer Reactions.*[21] The removal of an electron from the radical anion of a polynuclear aromatic (carbocyclic and heterocylic) or, more rarely, the addition of an electron to the radical cation often results in the production of the neutral substrate in an electronically excited state from which emission may occur. The electron transfers may be induced by oxidizing or reducing agents. Thus, oxidation of the radical anion of 9,10-diphenylanthracene (DPA) by 9,10-dichloro-9,10-diphenylanthracene ($DPACl_2$) induces an emission identical with that of DPA. A possible mechanism is:

$$\begin{aligned} DPA^{\dot{-}} + DPACl_2 &\rightarrow DPA + Cl^- + DPACl^{\cdot} \\ DPACl^{\cdot} + DPA^{\dot{-}} &\rightarrow DPA + Cl^- + DPA^* \\ DPA^* &\rightarrow DPA + h\nu \end{aligned} \qquad (2.17)$$

In molecular orbital terms, the process can be represented diagrammatically as in Figure 2.26.

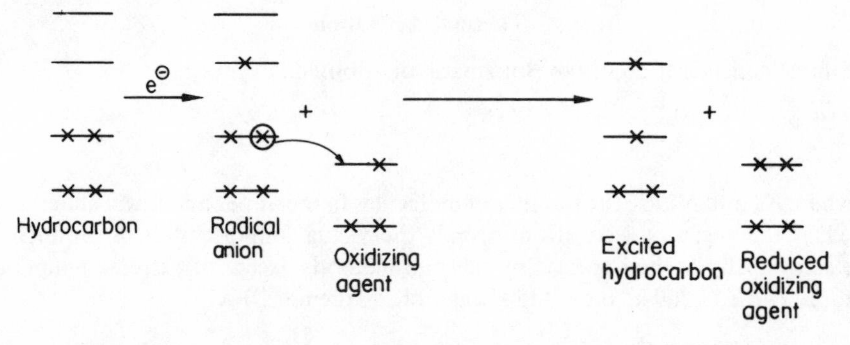

Figure 2.26. Energy level scheme for electron transfer chemiluminescence

The required electron transfers may be induced purely electrically. Many substances, dissolved in an aprotic solvent such as dimethylformamide containing tetra-n-butylammonium perchlorate as electrolyte, luminesce strongly when a few volts a.c. are impressed upon electrodes immersed in the solution, particularly when a fluorescer such as rubrene is present which may be excited by energy transfer from the substrate excited state. It should be emphasized that obscurity surrounds many of the details of the mechanism by which chemical energy is converted into radiation.

(b) *Singlet Oxygen.* When hydrogen peroxide is oxidized by chlorine in alkaline solution, a red glow is observed; this emission is from excited oxygen. The lowest excited states of the oxygen molecule are singlets $^1\Delta_g$ and $^1\Sigma_g$ (contrast the ground state, which is a triplet). These excited states have energies of 92 (22) and 160 kJ mol^{-1} (38 kcal mol^{-1}) respectively, and they are sufficiently energetic to emit in the red. It seems that it is a dimer of the $^1\Delta_g$ state (E = 185 kJ mol^{-1} (44 kcal mol^{-1}) ≡ 640 nm) which is the actual emitter. Equally, the dimer ($^1\Delta_g + {}^1\Sigma_g$) emits at 478 nm. Since the dimer $(^1\Sigma_g)_2$ would have 320 kJ mol^{-1} (76 kcal mol^{-1}) available energy, Khan and Kasha[22] advanced a general theory of chemiluminescence which proposed that energy transfer to suitable fluorescent species might be responsible for the chemiluminescence of those systems in which singlet oxygen is generated.

(c) *Peroxide Decompositions.* The most efficient (brilliant) chemiluminescent reactions involve the decomposition of peroxides, and many of these processes have been shown to involve the concerted and highly exothermic decomposition of intermediate dioxetanes (McCapra mechanism[23]). Tetramethyldioxetane was found to be weakly chemiluminescent on heating and much more so in the presence of a fluorescer excited by energy transfer (equation 2.18).

$$\text{O--O (tetramethyldioxetane)} \xrightarrow{\Delta} \text{O} + \text{O}^{*} \xrightarrow{\text{fluorescer}} \text{fluorescer}^{*} \longrightarrow h\nu \qquad (2.18)$$

Similarly, indolenyl peroxides glow with a beautiful green light on treatment with potassium t-butoxide in aprotic solvents, probably by the process (2.19).[24] The emission is identical with the fluorescence emission spectrum of the product.

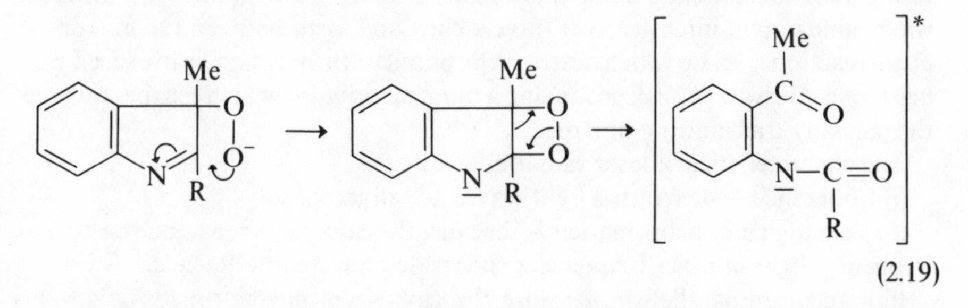

(2.19)

There are numerous other examples of chemiluminescent reactions which depend on intermediate dioxetane formation.

2.4 LASERS

It seems appropriate, at this point, to discuss lasers. A laser (*L*ight *A*mplification by *S*timulated *E*mission of *R*adiation), as is implied by its name, depends for its operation on stimulated emission of radiation—contrast conventional sources which function through spontaneous emission. Recalling (p. 12) that the rates of absorption and of stimulated emission for a given radiation density ρ are $n_l B_{lu}\rho$ and $n_u B_{ul}\rho$, where n_l and n_u are the number of molecules in the lower and upper states and B_{ul} and B_{lu} are the Einstein coefficients, it can be seen that if spontaneous emission could be suppressed in some way, a system exposed to radiation at the transition frequency would absorb radiation until $n_l = n_u$, i.e. until the ground state and excited state were equally populated. However, should a *population inversion* exist (i.e. if $n_l < n_u$), then light amplification can take place, for irradiation with light of the transition frequency will cause emission whose rate will be greater than the rate of absorption. This process will continue until $n_l = n_u$.

Lasers, which exploit this principle, comprise an emitting material set in a tuned optical cavity—a pair of parallel mirrors, one of which is partially transparent (Figure 2.27). The mirrors are separated by an integral number of

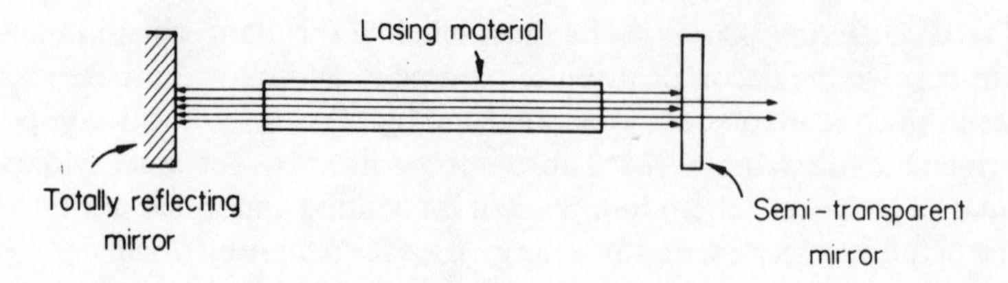

Figure 2.27. Schematic diagram of a laser

$\frac{1}{2}\lambda$ (where $E_u - E_l = hc/\lambda$). As a result, the light reflected from a mirror will be in phase with the incident wave (constructive interference). When a population inversion has been generated in the emitting material, *spontaneous* emission provides a few photons, which on collision with molecules of the excited lasing material stimulate it to emit in phase with the incident photons. Thus the wave builds up in intensity as it travels back and forth between the mirrors. A chain reaction is set up which destroys the population inversion in an exceedingly brief time interval (~ 1 μs), producing a burst of radiation which escapes through the partially transmitting mirror.

The characteristics of laser radiation are:

(i) Coherence—the emitted light waves are all in phase.

(ii) Very high monochromaticity—because the cavity is tuned to one particular frequency, light of other frequencies suffers destructive interference.

(iii) Accurate parallelism, because the cavity will not be tuned for off-axis radiation, which, in any case, would escape after only a few reflections. This lack of divergence of the beam permits it to be focussed to a spot of very small dimensions (of the order of a wavelength), thus generating very high radiation densities ($> 10^9$ W cm^{-2}) for short laser pulses.

(iv) Enormous brilliance of a laser pulse—a 'long' pulse of 1 μs containing 1 J of energy will have a power output of 1 MW, a figure which can be increased by orders of magnitude by shortening the laser pulse duration.

The population inversion essential to laser action cannot be achieved by direct radiative excitation to the lasing state, since this can at best cause n_u to equal n_l. Radiationless transitions or energy transfer must be employed in order to gain access to some excited state other than the one being populated by the energy input. For example, in the well-known ruby laser the lasing material is a cylindrical ruby rod, the Cr^{3+} ions of which are excited to the $^4T_{2g}$ state by an intense flash of light from a flash tube surrounding the rod. Rapid intersystem crossing populates the ^2E_g state from which the characteristic red laser emission occurs (Figure 2.28). Such a three-level system is inefficient—it can only work if more than 50% of the Cr^{3+} ions are excited to the ^2E_g state during the exciting flash. More efficient are the four-level systems (Figure 2.29) in which laser emission results from transition not to the ground state but to an unpopulated state C. If the state C, which is populated by the transition $B \rightarrow C$, is also rapidly depopulated by a fast conversion to the ground state, then it is possible to maintain a continuous population inversion of B with respect to C.

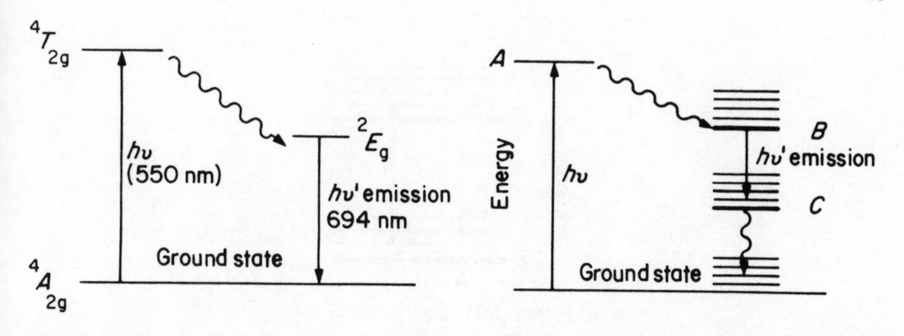

Figure 2.28. Energy levels for Cr^{3+} ions in ruby laser

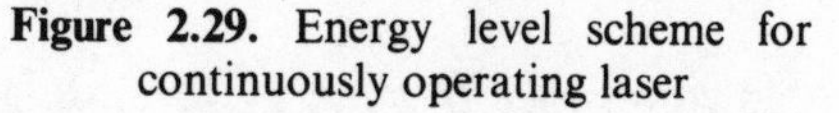

Figure 2.29. Energy level scheme for continuously operating laser

Such lasers will give a continuous light output. Some continuously operating gas lasers operate on similar principles. In the He–Ne laser, He is excited by an electrical discharge to the state *A* in Figure 2.29. Energy transfer affords Ne* (*B* in the Figure), which is the emitting species.

Similarly, in the CO_2 laser, N_2 is excited by electron impact into its first vibrational level. *Vibrational* energy transfer selectively populates the first vibrational level of the asymmetrical stretching mode of CO_2 with which it is nearly isoenergetic. The first vibrational level of the symmetric stretching mode is of lower energy and cannot be populated by this process. Lasing occurs through transitions between the rotational sublevels of these two states, giving rise to intense infrared emissions, particularly at 10·57, 10·59, and 10·61 μm.

Recently it has been shown that solutions of many strongly-fluorescent organic dyes can be induced to lase. The importance of these dye lasers[25] is that they can be tuned, i.e. the wavelength of their light output can be varied (often over as much as 50 nm). One type of apparatus, shown diagrammatically in Figure 2.30, replaces the totally reflecting plane mirror by a diffraction grating which returns to the cavity only light of a particular frequency determined by its angle of rotation.

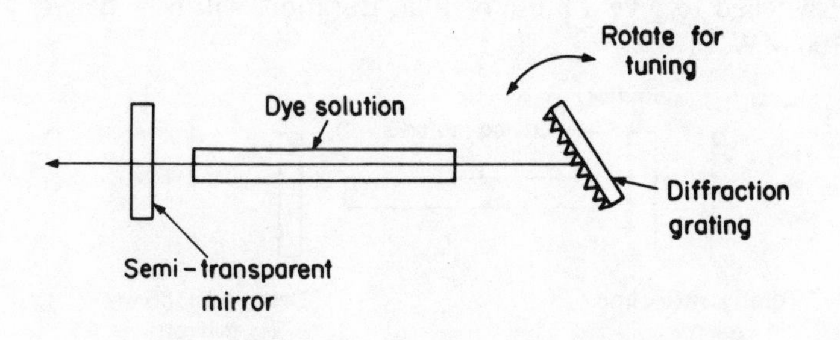

Figure 2.30. Schematic diagram of a tuneable dye laser

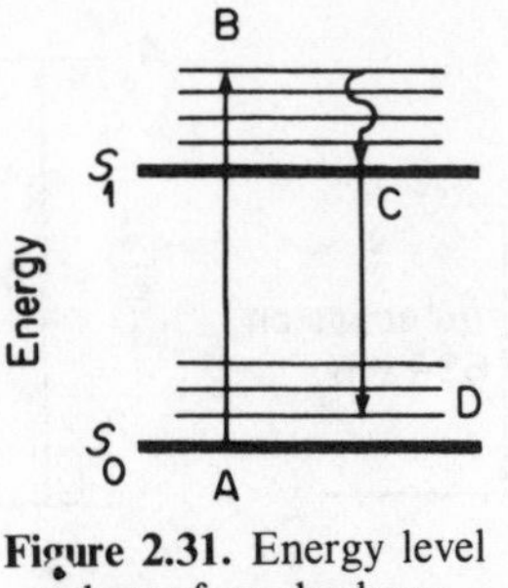

Figure 2.31. Energy level scheme for a dye laser

Now consider the energy level diagram for the lasing dyestuff (Figure 2.31). Pumping and vibrational relaxation promotes molecules of the dye to point C, the bottom vibrational level of S_1, from which they can radiatively collapse to the quasi-continuum of densely packed vibrational and rotational levels of S_0. If the cavity is tuned to a particular frequency corresponding, say, to the transition $C \rightarrow D$, then the laser will emit light of this frequency only. A moment's consideration shows that although only two electronic states are involved dye lasers are, in fact, four-level systems and as such may be made to operate continuously. Up to 70 % of the energy trapped in the S_1 state can be obtained in the form of laser emission at the prescribed frequency.

Laser Pulse Width[25b]

The simple laser (Figure 2.27) gives relatively long pulses ($\sim 1\ \mu s$) because spontaneous emission starts immediately upon excitation, but laser emission cannot begin until a population inversion has been achieved. The pulse width may be reduced to ~ 10 ns by the technique of Q-switching, which involves the interposition between the end of the laser rod and the totally reflecting mirror a shutter which can be opened with extreme rapidity (e.g. a Kerr cell, see Figure 2.32). If the laser rod is pumped with the shutter closed, lasing is impossible because light leaving the rod cannot get back to stimulate further emission (the amplification factor of the cavity is $\ll 1$). If the shutter is now opened, the amplification factor of the cavity suddenly rises to a value far beyond unity, and the stored energy is released as a giant pulse. A small ruby laser storing 1 J of energy, Q-switched to give a pulse of 10 ns duration, will now deliver a peak power of 100 MW.

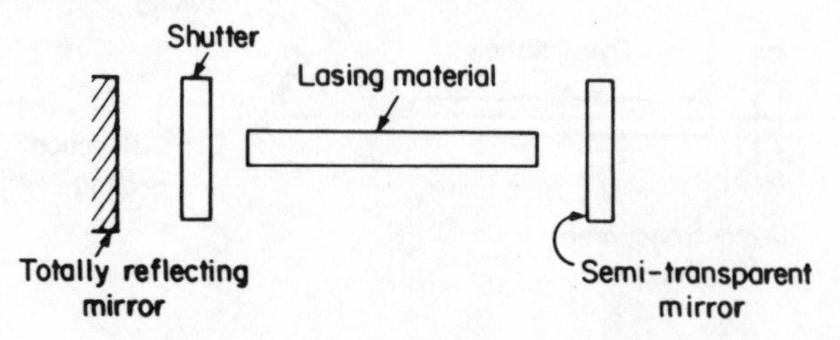

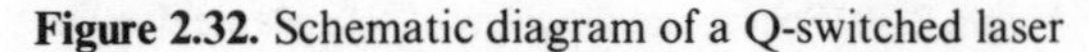
Figure 2.32. Schematic diagram of a Q-switched laser

Slight modification of the system to include the use of a dilute solution of a dye as the shutter results in a laser output consisting of a train of pulses ('mode-locking') having a half width of $1/\Delta\nu$, where $\Delta\nu$ is the half-width of the fluorescent emission of the laser material (Figure 2.33). By these means, pulses with pulse widths as short as 0·4 ps ($0{\cdot}4 \times 10^{-12}$ s) have been obtained with peak powers of 5 GW (to comprehend such short time intervals, remember that light requires 3 ps to travel 1 mm).

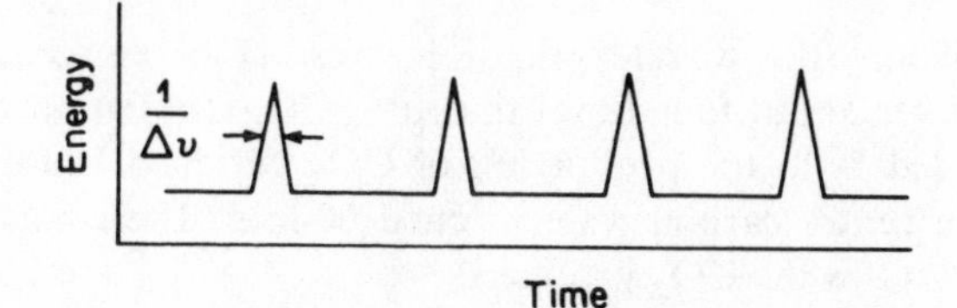

Figure 2.33. Pulse output of a mode-locked laser

2.4.1 Laser Applications[26]

Energy transfer, intersystem crossing and vibrational relaxation are extremely fast processes. Their rates can now be measured directly using picosecond flashes from a mode-locked laser (see Chapter 3, p. 75). Also the very narrow line-width of laser emission means that the output from a dye laser can be tuned to excite a specific vibrational level of a molecule in the gas-phase in order to study its decay characteristics.

Forbidden processes are characterized by very low extinction coefficients, so that only a small fraction of the incident light can be absorbed and used to effect these processes. The enormous intensities of laser pulses overcome this difficulty. For example, the photon scattering giving rise to the Raman effect is extremely weak—only about 10^{-6} of the incident light appears as Raman scattering. The very high power output per unit area of a focussed laser makes it an ideal light source and permits the use of very much smaller sample specimens. Furthermore, at the focussed output of a giant pulse laser, novel phenomena are observed—the stimulated Raman effect and the inverse and hyper-Raman effects, which are excellently summarized by Long.[27] Another forbidden process is multiphoton absorption. This was first demonstrated[28] through the blue fluorescence ($\lambda = 425$ nm) of Eu^{2+} ions when subjected to the focussed red output of a pulsed ruby laser ($\lambda = 694{\cdot}3$ nm). The phenomenon is due to the 'simultaneous' absorption of two photons. A recent application of this phenomenon is two-photon spectroscopy,[29] a simple procedure for recording the population of otherwise inaccessible ultraviolet vibronic levels using the visible light output from a tunable dye laser.

Replacement of an atom in a molecule by its isotope induces small shifts of the molecular electronic, vibrational and rotational levels. If, as a result of these isotopic shifts, there is a particular frequency at which one species absorbs and the other is transparent, then excitation at that frequency will lead to selective excitation of one isotopic component. This is the basis of the rapidly developing field of laser isotope enrichment.[26,28] For the process to be feasible certain conditions must be fulfilled: (i) the exciting light must be of sufficiently narrow bandwidth to excite just one of the isotopic species (this implies laser or atomic emission); (ii) energy transfer between the different isotopic components must be minimized; (iii) the light absorption must lead to a different species to permit separation. A few examples will serve to illustrate the technique:

(1) At 123·58 nm (the wavelength of Kr resonance emission) the optical density of ^{13}CO is about four times that of ^{12}CO. Irradiation of natural CO with this light led[28c] to the production of CO_2 enriched 10-fold in ^{13}C, and of C_3O_2 whose centre carbon was enriched 60-fold. There was also a 7-fold enrichment of ^{18}O in the CO_2 produced.

(2) The ^{235}U in uranium vapour can be selectively excited by a laser beam and then photoionized by another. The $^{235}U^+$ is removed from unaffected ^{238}U by electric and magnetic fields.

(3) Multiphoton absorption by SF_6 of one of the CO_2 laser lines leads to dissociation of an S—F bond and to a 33-fold change in the $^{34}S:^{32}S$ ratio in the unreacted SF_6. This process required the simultaneous absorption of numerous infrared photons, and it is only possible with enormous power densities ($\sim$6 GW cm^{-2}) at the focussed output of a large pulsed CO_2 laser.

2.5 PROPERTIES OF EXCITED STATES

Excited species are subject to rate processes such as emission, quenching, etc. They also have a range of static properties which are independent of time and are not the consequence of the conversion of the species into another entity. It is with these latter properties that we are now concerned. Dynamic properties will be discussed in the next chapter.

2.5.1 Geometry of Excited Molecules

The impossibility of obtaining over a sufficient length of time a sufficiently high concentration of excited species to be examined by the usual techniques of structural analysis means that the geometries of excited molecules are extremely difficult to determine. The analysis of vibrational and rotational fine structure of high resolution absorption spectra obtained in the gas phase is one of the few methods available, and provides information about the vibrational modes and rotational constants. Another technique can be applied to transitions which are symmetry forbidden. If the molecule in its ground state has a vibration which transforms its shape towards that of the excited state, then the corresponding vibronic band appears strongly in the gas-phase spectrum. Identification of strong vibronic transitions then provides valuable clues to the structure of the excited state. Such techniques, which can only be applied to very simple molecules, show that excitation can lead to profound changes in geometry. The results can be qualitatively rationalized by simple molecular orbital considerations and supported by numerical calculations.

Acetylene[30]

The ultraviolet absorption band of acetylene at 250–210 nm corresponds to a singlet ($\pi \rightarrow \pi^*$) transition ($^1\Sigma_g^+ \rightarrow {}^1A_u$) to an excited state which has the transoid planar trigonal structure of Figure 2.34, in which the pair of electrons originally in one of the π orbitals now occupy sp^2 orbitals on the carbon atoms. The transition is symmetry forbidden and is further weakened ($f \sim 10^{-4}$) because the change in molecular architecture causes such a large relative displacement of the potential energy surfaces of the ground state and excited state that the vibrational overlap integral is very small.

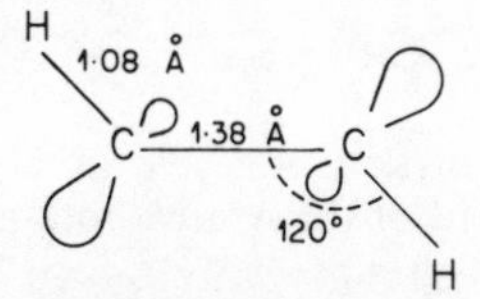

Figure 2.34. Geometry of excited acetylene (^1A_u)

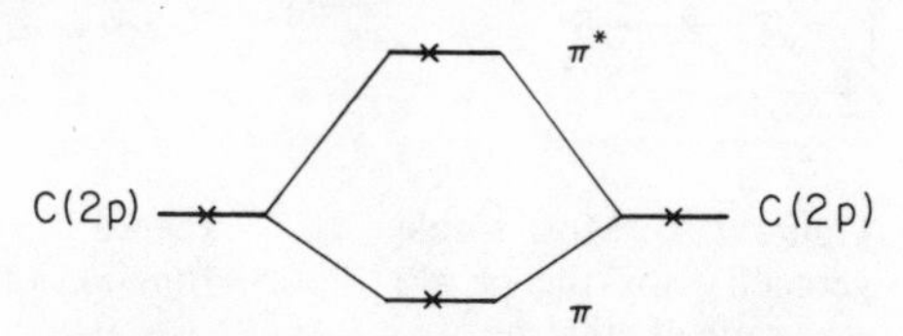

Figure 2.35. In a double bond, the π^* level is more antibonding than the π level is bonding

Reference to Figure 2.35 illustrates that the antibonding induced by the electron in the π^*-orbital is greater than the bonding of the single electron in the π-orbital. In other words, stability may be obtained by uncoupling the π and π^*-bond and 'locating' the two electrons in C 2p orbitals, when further stability may be attained by rehybridization to sp^2. In valence bond terms this excited state could be crudely represented as a resonance hybrid:

$$\mathrm{H\dot{C}{=}\dot{C}H} \leftrightarrow \mathrm{H\overset{+}{C}{=}\overset{-}{C}H} \leftrightarrow \mathrm{H\overset{-}{C}{=}\overset{+}{C}H}$$

It must be emphasized that the electronic transition from linear ground state acetylene gives rise (Franck–Condon principle) to a linear excited state. The relaxation to the more stable bent state occurs after excitation.

Ethylene[31]

The most stable structure for the lowest energy (π, π^*) state of ethylene appears from both experiment and calculation to be one in which the two CH_2 groups are joined by what is essentially a single bond and lie in perpendicular planes (Figure 2.36) with two electrons occupying p-orbitals on the carbon atoms. The data can be rationalized by noting that, as with excited acetylene, the promotion of an electron to a π^*-orbital leads to a system which becomes more stable by uncoupling the π-orbital to leave only a σ-bond between the carbon atoms. The two electrons, now in the C 2p orbitals, experience electrostatic repulsion, and this is readily minimized by rotation about the C—C single bond (Figure 2.37). Presumably the carbon atoms of the two orthogonal CH_2 groups rehybridize to a pyramidal structure in order to incorporate stabilizing

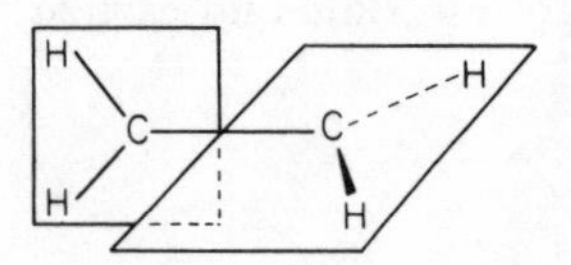

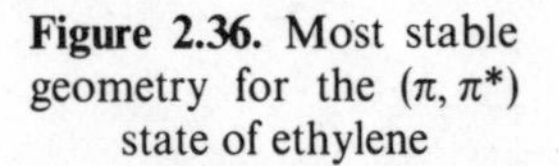

Figure 2.36. Most stable geometry for the (π, π^*) state of ethylene

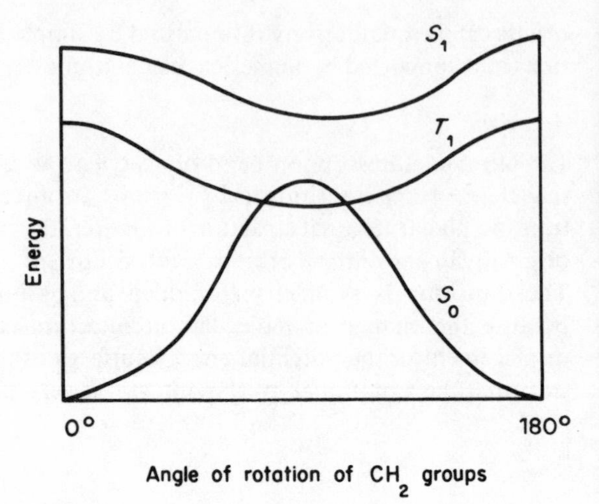

Figure 2.37. Energy of the ethylene molecule as a function of the relative rotations of the relative rotations of the CH_2 groups

's' character into the orbitals of the two electrons. Similar considerations apply to the triplet (π, π^*) state of ethylene.

Again, the perpendicular state is not the Franck–Condon state obtained immediately upon excitation, and it is referred to as a *non-vertical* or *non-spectroscopic* state. The rotations consequent upon excitation of ethylene presumably also occur in substituted olefins, and they are thought to constitute the underlying basis for *cis–trans* photoisomerization.

Formaldehyde

The important difference between the (n, π^*) state of formaldehyde and the (π, π^*) states of ethylene and acetylene is that in the former there are *two* electrons in the bonding π-orbital. The antibonding induced by the single electron in the π^*-orbital is inadequate to neutralize completely the bonding power of the doubly occupied π-orbital. However, it does weaken the bond and hence lengthen it to a value intermediate between those for C—O and C=O. Furthermore, the electron in the π^*-orbital increases the electron density at the carbon

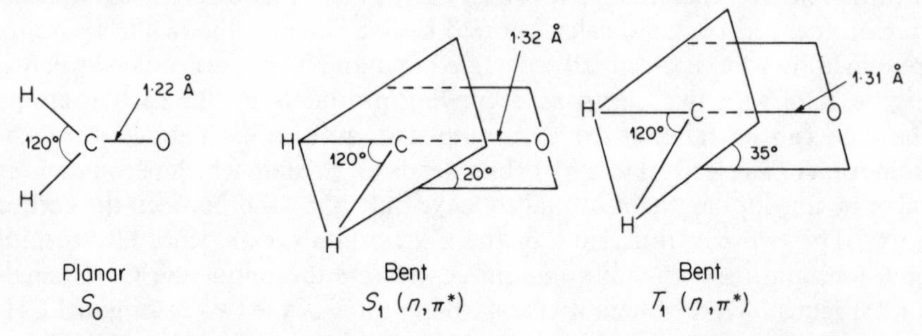

Figure 2.38. Geometries of the lowest electronic states of formaldehyde

atom. As a result, this atom rehybridizes in order to confer some ‘s’ character upon the promoted electron. This implies loss of coplanarity (Figure 2.38). The experimental data[32] support these expectations.

It is to be expected that in more extended chromophores the geometrical changes consequent upon excitation will be much smaller. In the lowest (π, π^*) state of ethylene, 50% of the π-electrons are promoted. In naphthalene, for example, only 1 π-electron in 10 would be involved in the $(\pi \rightarrow \pi^*)$ transition.

2.5.2 Acid–Base Properties[33]

It has been known for some time that the fluorescence spectra of certain phenols and aromatic amines are pH-dependent. In Figure 2.39 curves 1 and 5 are the fluorescent emissions of 2-naphthoxide and 2-naphthol respectively. At intermediate pH (curves 2, 3 and 4) emission is observed from both 2-naphthol and 2-naphthoxide even though at these acid concentrations there is no significant amount of the anion present in the solution being scanned. Förster explained the phenomenon by postulating that proton exchanges are so rapid that an acid–base equilibrium is established (2.20) between the excited phenol (HA^*) and its *excited* conjugate base (A^{-*}):

$$\begin{array}{ccccc} HA \xrightarrow{h\nu} & HA^* & \rightleftarrows & H^+ + & A^{-*} \\ & \downarrow & & & \downarrow \\ & HA + h\nu' & & & A^- + h\nu'' \end{array} \tag{2.20}$$

The implication of Figure 2.39 is that the phenols in their S_1 states are much stronger acids than in their ground states.

Two methods have been employed to measure the pK_a values of the excited singlets of phenols and amines. Both assume proton equilibrium during the lifetime of the excited state (10^{-8} to 10^{-9} s).

(i) The pK_a is that pH at which the fluorescence of HA^* drops to one-half of its intensity in solutions so strongly acid that only HA^* emits. It is also the pH at which the intensity of the fluorescence of A^{-*} drops to one-half of the value observed in solutions so alkaline that only A^{-*} emits.

(ii) The second method depends on the Förster–Weller cycle (Figure 2.40). It is clear from the diagram that:

$$\Delta E_{HA} - \Delta E_{A^-} = \Delta H - \Delta H^* = \Delta G - \Delta G^* \tag{2.21}$$

assuming that the entropy of dissociation is the same in both ground state and excited state. It follows that

$$\ln\left(\frac{K^*}{K}\right) = \frac{\Delta E_{HA} - \Delta E_{A^-}}{RT} = \frac{h\,\Delta\nu}{kT} \tag{2.22}$$

where K^* and K are the dissociation constants of HA^* and HA, and $\Delta\nu$ is the difference in the frequency of absorption or emission of HA and A^- measured in the (O—O) bands. Hence K^* may be obtained. Similarly, the pK_a values of triplet states have been estimated by comparing the absorption and phosphorescent emission spectra of triplet species. Some values of pK_a for deprotonation are collected in Table 2.1.

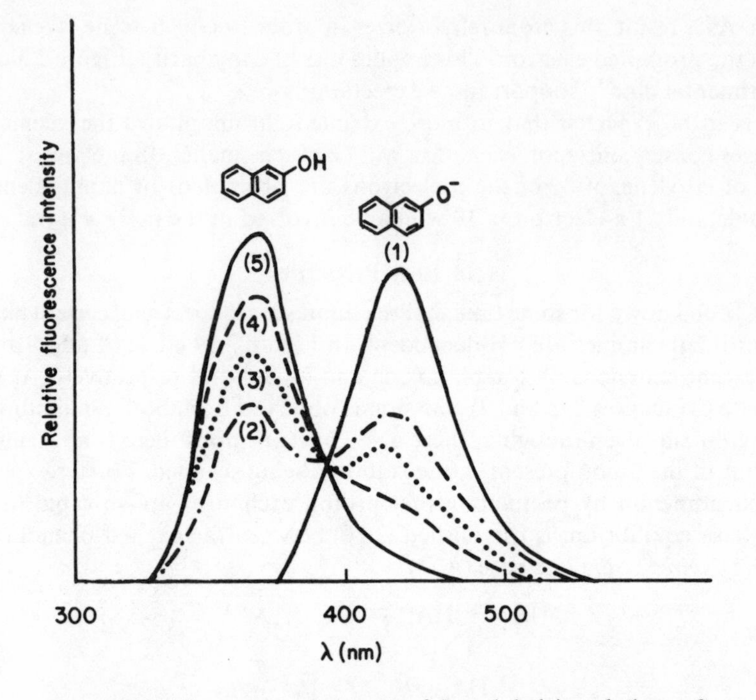

Figure 2.39. Fluorescence spectrum of 2-naphthol in solutions of different pH. (1) 0·02 M NaOH; (2) 0·02 M acetic acid + 0·02 M sodium acetate; (3) pH 5–6; (4) 0·004 M $HClO_4$; (5) 0·15 M $HClO_4$ (from A. Kearwell and F. Wilkinson, in G. M. Burnett and A. M. North (ed.), *Transfer and Storage of Energy by Molecules*, volume 1, (1969), John Wiley and Sons Ltd.)

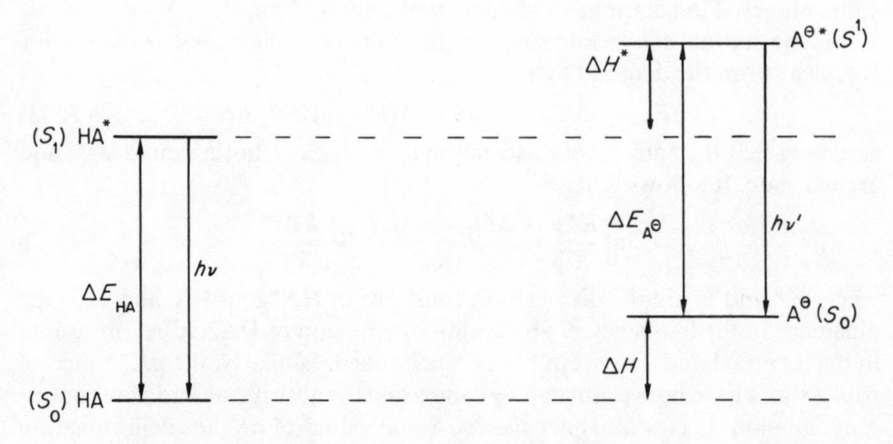

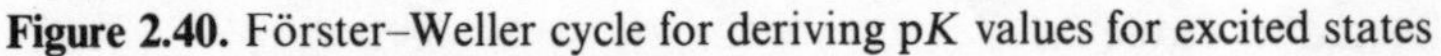
Figure 2.40. Förster–Weller cycle for deriving p*K* values for excited states

Table 2.1. pK_a values for deprotonation of ground and excited states

Molecule	p$K_a(S_0)$	p$K_a(S_1)$[a]	p$K_a(T_1)$[a]
p-Cresol	10·3	4·1–5·7	8·6
1-Naphthol	9·2	2·0	
2-Naphthol	9·5	2·5–3·4	7·7–8·1
2-Naphthylamine salts	4·1	−1·5– −2·9	3·1–3·3
1-Naphthoic Acid	3·7	10–12	3·8–4·6
Acridine salts	5·5	10·6	5·6
Indole	very large	12·3	

[a] The values obtained depend upon the method used for their estimation.

Examination of the data in the Table reveals the following features:

(i) For phenols and the salts of aromatic amines, the lowest *singlet* excited state is much more acidic than the ground state by a factor of 10^5–10^6.

(ii) Conversely, aromatic acids and the conjugate acids of certain heterocyclic amines are much weaker acids in the singlet excited state than in the ground state.

(iii) The pK_a values of excited triplet states are comparable with those of the ground states. This phenomenon is perhaps most readily explained in valence bond terms. The wavefunctions for S_0 and S_1 states of phenol can be approximated by mixing the wavefunctions corresponding to the structures **I**, **II** and **III** (2.23).

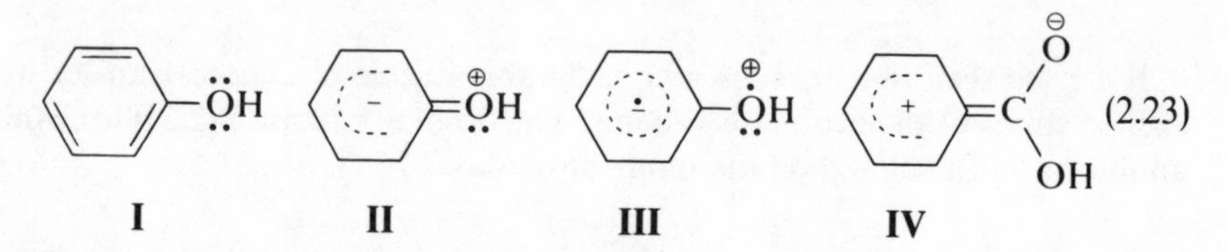

The important difference in the two states lies in the greatly enhanced contribution of structure **II** to the resonance hybrid constituting S_1, which leads to increased positive charge on the oxygen atom and hence to greatly increased acidity of the excited state. Structure **II**, which implies the transfer of a pair of electrons to the aromatic nucleus, will not contribute significantly to the T_1 state because spin correlation tends to keep electrons with parallel spins apart (see discussion, p. 54). Hence the oxygen atom is less positively charged in the T_1 state than in the S_1 state, leading to the observed sequence p$K_a(S_1)$ < p$K_a(T_1)$ < p$K_a(S_0)$. A similar argument explains the increased acidity and diminished basicity of aromatic amines.

In aromatic carboxylic acids, excitation to S_1 leads to charge migration away from the ring, something which can be represented by an increased contribution of the charged structure **IV** (2.23) to the resonance hybrid constituting S_1. The effect will be to reduce the acidity of the S_1 states of aromatic carboxylic acids and greatly enhance their basicity.

2.5.3 Dipole Moments[34]

Because electronic transitions lead to redistribution of electrons, the dipole moments of a molecule in its ground state and excited state will, in general, be different. Estimates of the dipole moments of excited states may be obtained by analysing the effect of solvents on the absorption and fluorescence maxima,[35] and more directly by Czekalla's method.[36] This depends on the application of an electric field to a solution being irradiated with plane polarized light. The molecules in both their ground and excited states tend to align themselves with the field to an extent determined by their dipole moment, and this is manifested as a change in the degree of polarization (see Chapter 3, p. 76) of the fluorescent emission. Table 2.2 summarizes some of the data.

Table 2.2. Dipole moments of some singlet states

Compound	Dipole moment (in Debyes) S_0	S_1[a]
p-Nitroaniline	6	14
4-Amino-4′-nitrobiphenyl	6	18–23
2-Amino-7-nitrofluorene	7	19–25
4-Dimethylamino-4′-nitrostilbene	7·6	20·7–32
Formaldehyde	2·3	1·6

[a] The values obtained depend on the method adopted for their measurement.

It can be seen that there is a considerable degree of charge–transfer in the excited states of the aromatic systems. The dipolar structure (2.24) for *p*-nitroaniline has a calculated dipole moment of about 25 D.

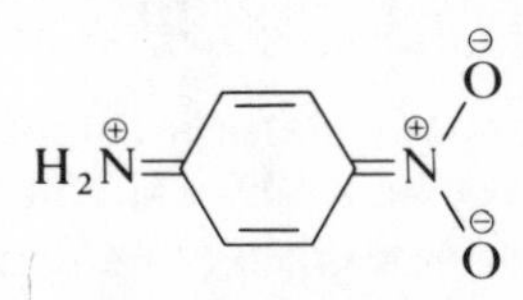

(2.24)

2.6 ENERGIES OF EXCITED STATES

The energy of a singlet or triplet excited state (E_S or E_T) is normally taken to be the energy difference between the $v = 0$ levels of the excited state and the corresponding ground state. It is therefore a quantity determinable spectroscopically by locating the 0–0 transition. If vibrational fine structure is present, the measurement is readily made for S_1 and T_1 states by recording the $S_0 \rightarrow S_1$ and $S_0 \rightarrow T_1$ absorption and/or emission spectra. In the absence of vibrational structure, the energies of the states may be estimated roughly from the position of the long wavelength tail of the absorption spectrum, recorded at low temperature to eliminate 'hot' bands.

The precise determination of the energies of states other than the lowest singlet and triplet is more difficult. Emission spectra are useless because emission almost invariably occurs from the bottom state; $S_0 \rightarrow S_n$ transitions usually lack fine structure and tend to merge into the short wavelength $S_0 \rightarrow S_1$ absorption. The same is true for upper triplets. Fortunately, for most purposes, it is the energy of the lowest states which are of greatest interest to photochemists.

2.6.1 Singlet–Triplet Splitting

The Table 2.3 records E_{S_1} and E_{T_1} for some representative systems. It is apparent that in all cases $E_{T_1} < E_{S_1}$. This result is quite general—all triplets are more stable than the corresponding singlets (i.e., singlets with the same orbital

Table 2.3. Singlet–triplet splittings of excited states

Compound	E_{S_1} in kJ (kcal) mol^{-1}	E_{T_1} in kJ (kcal) mol^{-1}	Singlet–Triplet splitting in kJ (kcal) mol^{-1}	Transition
Formaldehyde	338 (80·6)	302 (72·1)	36 (8·5)	$n \rightarrow \pi^*$
Acrolein	310 (74·0)	291 (69·3)	19 (4·7)	$n \rightarrow \pi^*$
Benzophenone	318 (75·8)	291 (69·4)	27 (6·4)	$n \rightarrow \pi^*$
Benzene	461 (110)	354 (84·3)	107 (25.5)	$\pi \rightarrow \pi^*$
Naphthalene	386 (92·1)	255 (60·8)	131 (31.2)	$\pi \rightarrow \pi^*$
Tricyclic, catacondensed aromatic hydrocarbons[a]			~132 (31)	$\pi \rightarrow \pi^*$
Tetracyclic, catacondensed aromatic hydrocarbons[a]			~125 (30)	$\pi \rightarrow \pi^*$
Heptacyclic, catacondensed aromatic hydrocarbons[a]			~114 (27)	$\pi \rightarrow \pi^*$
Ethylene		344 (82·1)	290 (69·2)[c]	$\pi \rightarrow \pi^*$
Butadiene		250 (59·6)	271 (64·6)[c]	$\pi \rightarrow \pi^*$
Toluene/tetra-cyanobenzene	288 (68·6)[b]	252 (60·1)[b]	36 (8·5)	Charge–transfer
Hexamethylbenzene/tetracyanobenzene	222 (52·9)[b]	206 (49·1)[b]	16 (3·8)	Charge–transfer

[a] In catacondensed hydrocarbons, no carbon atom lies at the junction of more than 2 rings.
[b] Positions of emission maxima.
[c] From oxygen perturbation of absorption spectra.

configuration). The energy difference ($E_S - E_T$) is known as the *singlet–triplet splitting* and has large values for (π, π^*) states and small ones for (n, π^*) and CT states. These points require explanation.

Spatial wavefunctions must be either symmetric Ψ_S or antisymmetric Ψ_A with respect to the interchange of the coordinates of two electrons. The antisymmetric wavefunction for a system of two electrons can be written

$$\Psi_A = \frac{1}{\sqrt{2}}[\psi(r_1)\psi'(r_2) - \psi(r_2)\psi'(r_1)]$$

where ψ and ψ' are the space orbitals, and r_1 and r_2 are the coordinates of the electrons. Clearly such a wavefunction is antisymmetric, because interchanging the electrons gives

$$\frac{1}{\sqrt{2}}[\psi(r_2)\psi'(r_1) - \psi(r_1)\psi'(r_2)] = -\Psi_A$$

Ψ_A becomes zero when $r_1 = r_2$, and since wavefunctions must be continuous a plot of the variation of Ψ_A against $(r_1 - r_2)$ is of the form shown in Figure 2.41.

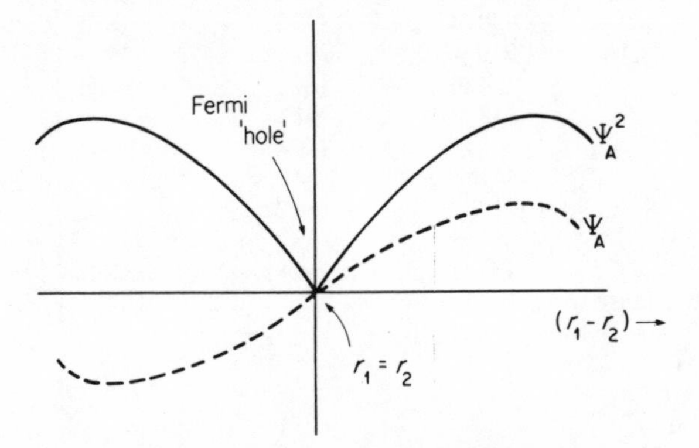

Figure 2.41. Fermi hole associated with an antisymmetric spatial wavefunction

The probability of finding one electron at r_1 and the other at r_2 is given by Ψ_A^2, and it is apparent that the probability of the electrons being in close proximity is very small (*Fermi hole*), reducing to zero when $r_1 = r_2$ and they are at the same point in space. For the symmetric wavefunction Ψ_S there is no Fermi hole. Therefore, on average, the electrons are further apart in systems described by Ψ_A than in those described by Ψ_S, with a corresponding reduction of Coulombic repulsion. Hence the energies of the spatially antisymmetric states Ψ_A will always be less than those of the corresponding symmetric states Ψ_S.

We now consider electron spin; this appears in wavefunctions as either α or β. Thus for two electrons the possible spin combinations are $\alpha(1)\alpha(2)$, $\beta(1)\beta(2)$, $\alpha(1)\beta(2)$ and $\beta(1)\alpha(2)$, of which the last two wavefunctions are unacceptable since they are neither symmetric nor antisymmetric with respect to electron interchange. They are therefore replaced by the linear combinations ($\alpha(1)\beta(2) \pm \alpha(2)\beta(1)$) which do satisfy the symmetry requirements. Three of these spin wavefunctions are symmetric and one is antisymmetric (2.25), and these are the wavefunctions for triplet and singlet states respectively. Since the Pauli Principle requires that the *total* electronic wavefunction (adequately represented as the product of space and spin functions) must be antisymmetric, it follows that triplet states will have an antisymmetric spatial wavefunction and singlets a symmetric one:

$$^1\Psi = \Psi_S \,.\, \Psi_A(\text{spin}); \qquad ^3\Psi = \Psi_A \,.\, \Psi_S(\text{spin})$$

Hence any triplet state will always be more stable than the corresponding singlet, simply because its space wavefunction is antisymmetric with the inevitable result that the charge centres of the two electron clouds are further apart than in the corresponding singlet.

triplet spin wavefunctions	singlet spin wavefunction
$\alpha(1)\alpha(2)$	
$\beta(1)\beta(2)$	
$\frac{1}{\sqrt{2}}[\alpha(1)\beta(2) + \beta(1)\alpha(2)]$	$\frac{1}{\sqrt{2}}[\alpha(1)\beta(2) - \beta(1)\alpha(2)]$ (2.25)
Symmetric	*Antisymmetric*

The singlet–triplet splitting, which is the difference in energy between singlet and triplet states, is thus a measure of the difference in electron distribution between singlet and triplet states. The situation just described implies that electrons with parallel spins avoid each other (*spin correlation*).

In molecular orbital theory, the singlet–triplet splitting is twice the *exchange integral* K_{ul} given by (2.26), where the subscripts u and l refer to upper and lower states concerned.†

$$K_{ul} = \iint \psi_u(1)\psi_l(1)\left(\frac{e^2}{r_1 - r_2}\right)\psi_u(2)\psi_l(2)\, dr_1\, dr_2 \qquad (2.26)$$

Recalling that the products $\psi_u\psi_l$ in this expression are the integrands of overlap integrals (p. 26), it can be deduced that if the two electrons are located in orbitals which overlap but slightly, the products $\psi_u\psi_l$ will be small and the singlet–triplet splitting will be small, as with (n, π^*) states and the excited states of donor–acceptor complexes. Similarly, in extended chromophores, where the orbitals involved in the transition are large, the average values of ψ_u and ψ_l will be smaller than in more compact systems, and this leads to a reduction in K_{ul} and the singlet–triplet splitting. These expectations are all borne out by the date in Table 2.3.

† Strictly, $K_{ul} = \iint \psi_u^*(1)\psi_l^*(1)\left(\frac{e^2}{r_1 - r_2}\right)\psi_u(2)\psi_l(2)\, dr_1\, dr_2$, but in this book the wavefunctions are assumed to be real (see footnote, p 15).

2.6.2 Singlets, Triplets and Biradicals[37]

The literature is replete with examples of excited states formulated as biradical structures. Thus the (π, π^*) state of ethylene is frequently written as (2.27), implying a high chemical reactivity of the carbon centres linked by the single bond. Similarly, the (n, π^*) state of formaldehyde is formulated as (2.28).

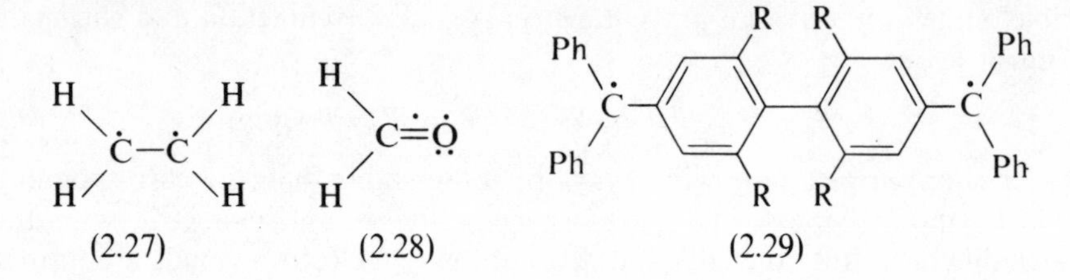

The wide dissemination of structures such as these has led to the belief in the biradical character of excited states. Whether excited states are biradicals or not is partly a matter of definition, but considerable added confusion has arisen because triplet states are paramagnetic (as are free radicals) but singlet states are not.

The problem may be clarified by first considering what is meant by a biradical. The ultimate biradical would be a *bifunctional* system in which the two radical centres behaved quite independently of each other. The e.s.r. spectrum of such a system would show a pair of doublets, since each electron could independently have the α or β spin. Such a situation, of which the tetrachloro derivative of Chichibabin's hydrocarbon (2.29), R = Cl) is a rare example, can be represented diagrammatically as in Figure 2.42.

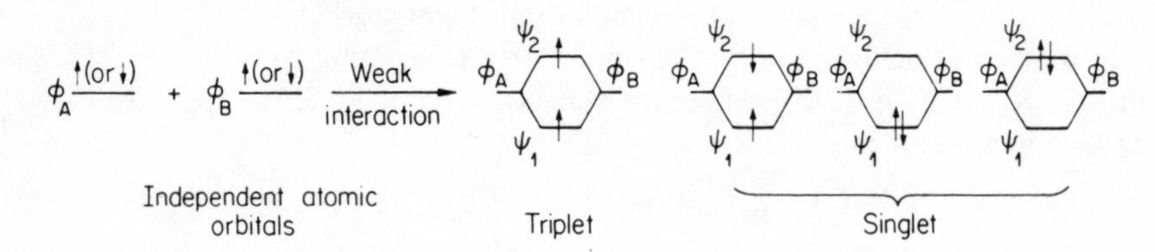

Figure 2.42. States derived from weakly interacting atomic orbitals

Case A

If, as is commonly the case, two degenerate or nearly degenerate radical centres with wavefunctions ϕ_A and ϕ_B interact weakly, then distinct molecular orbitals ψ_1 and ψ_2 embracing both centres develop. Since the interaction is weak, the energy difference $(E(\psi_1) - E(\psi_2))$ will be small, ψ_1 and ψ_2 will be almost degenerate, and the two electrons can be distributed over the two orbitals as shown in Figure 2.42. This gives a triplet biradical (detectable by e.s.r.) and a singlet biradical state whose wavefunction can only be properly described as an appropriate mixture of the three configurations in Figure 2.42.

Case B

Now suppose that the overlap between ϕ_A and ϕ_B (given by the overlap integral $S_{AB} = \langle\phi_A\phi_B\rangle$) increases. Then ψ_1 and ψ_2 separate further (Figure 2.43), the electrons therefore tend to pair off in ψ_1 for energetic reasons, and the singlet now acquires closed shell character and becomes stabler than the triplet. The singlet approximates increasingly to that state in which a covalent bond is formed between the two radical centres ϕ_A and ϕ_B.

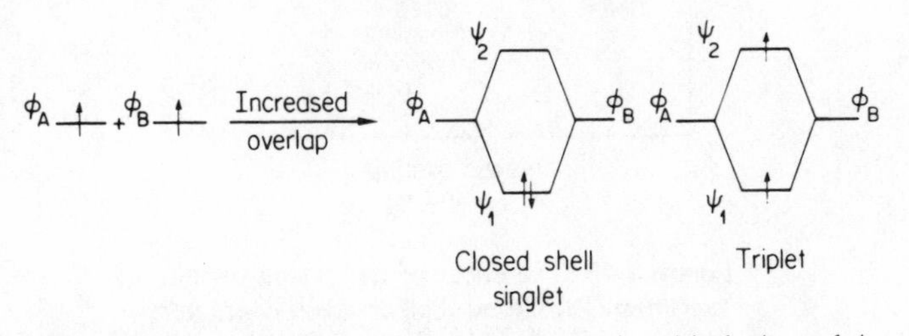

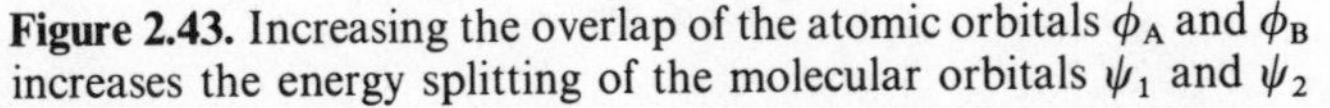

Figure 2.43. Increasing the overlap of the atomic orbitals ϕ_A and ϕ_B increases the energy splitting of the molecular orbitals ψ_1 and ψ_2

Case C

If ϕ_A and ϕ_B cease to be degenerate, then again ψ_1 and ψ_2 separate in energy (Figure 2.44), and ψ_1 approximates more and more closely to ϕ_A. Again the most stable state of the system is the singlet, but in this case it corresponds approximately to the zwitterionic state with both electrons located in ϕ_A.

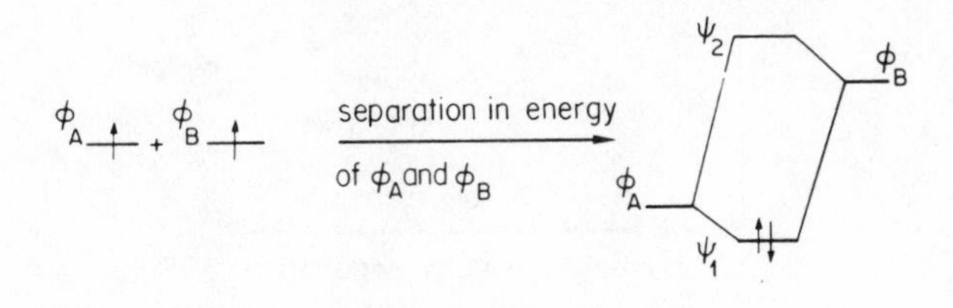

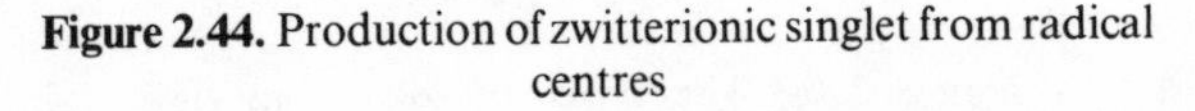

Figure 2.44. Production of zwitterionic singlet from radical centres

These trends are summarized in Figures 2.45 and 2.46.

If one takes the view just advanced that biradical character occurs only when the singlet–triplet splitting is small† and when the *molecular* orbitals are close in energy, then biradicals occur on the left of these diagrams and, proceeding through an ill-defined (shaded) region, arrive at an area on the right of the diagrams where the systems in their ground state acquire closed shell (covalent or zwitterionic) character.

† If the overlap between ϕ_A and ϕ_B is small then the exchange integral and therefore the singlet–triplet splitting are also small.

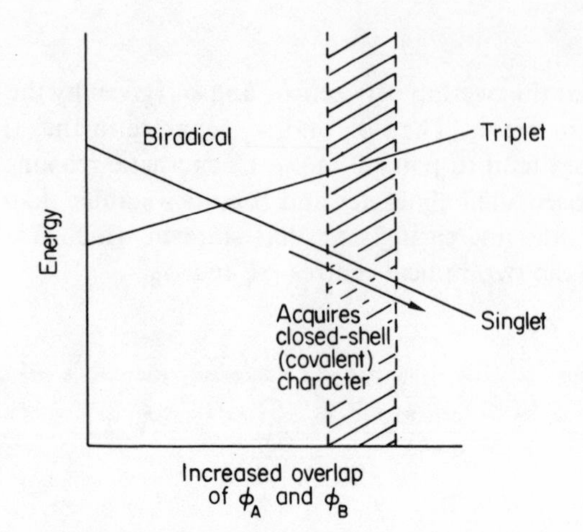

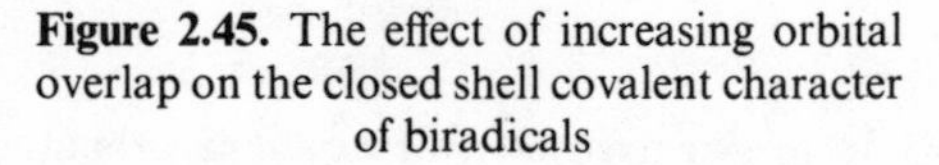
Figure 2.45. The effect of increasing orbital overlap on the closed shell covalent character of biradicals

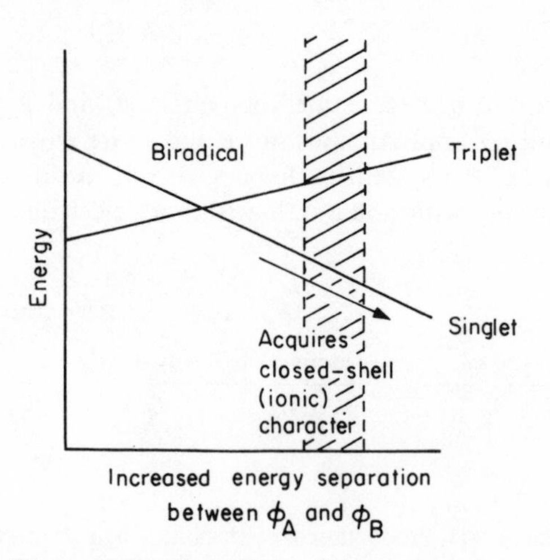

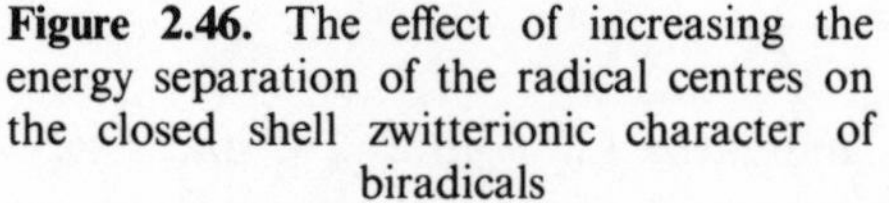
Figure 2.46. The effect of increasing the energy separation of the radical centres on the closed shell zwitterionic character of biradicals

On this basis the *planar* triplet and singlet excited states of ethylene would not qualify as biradicals because the large overlap between ϕ_A and ϕ_B means that the molecular orbitals ψ_1 and ψ_2 would have large energy separation. Consequently in the planar triplet state the electron in ψ_2 would be much more reactive than that in ψ_1, leading to radical but not *bi*radical behaviour. The singlet state would be expected to show little radical behaviour. On the other hand the relaxed orthogonal (π, π^*) excited states of ethylene (2.30), dienes (2.31) and T_1 benzene (2.32) would be classified as biradicals.

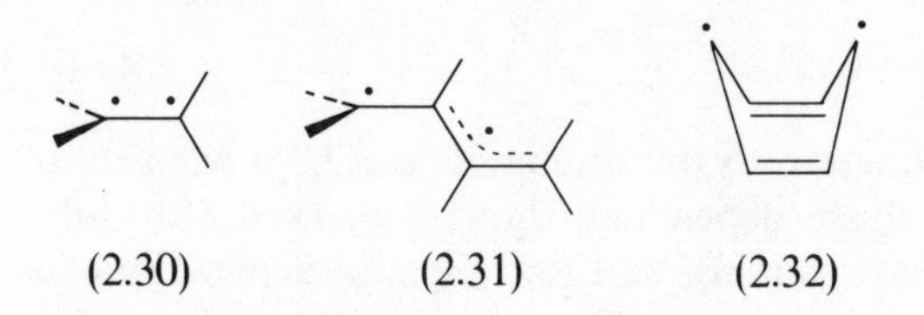

(2.30) (2.31) (2.32)

It should be noted that there is, as yet, no consensus of opinion on these matters and alternative views have been advanced. For example, it has been suggested that a divalent intermediate which reacts like a radical can probably be termed a biradical, independent of whether it is of singlet, or triplet multiplicity.

If S_1 and T_1 states of a molecule had similar electron distributions one would expect their chemical reactivity† to be similar except with regard to pericyclic reactions and processes where the operation of the Wigner spin conservation rules (see Chapter 4, p. 101) might impose a difference. However, the singlet–triplet splitting, the energy difference between the singlet and corresponding triplet states, is merely a measure of this difference in electron distribution (see Chapter 2, p. 55). Therefore, in (n, π^*) states, where the singlet–triplet splitting is small, the charge distributions will be closely similar. We therefore expect the two states to have similar photochemical behaviour. Conversely, where the singlet–triplet splitting is large, as in the (π, π^*) state of benzene, singlet and triplet photochemistry is expected to be dissimilar.

2.6.3 Solvent Effects

When absorption spectra are measured in solvents of increasing polarity it is found that for some systems λ_{max} moves to longer wavelengths (*red or bathochromic shift*) and for others an inverse *blue or hypsochromic* shift is encountered. These shifts provide information on the nature of the transition and can afford estimates of the dipole moments of excited states. This is a difficult area, requiring for its proper interpretation[38] consideration of solute–solvent interactions (dipole–dipole, dipole–polarization, hydrogen bonding) and of the impact of the Franck–Condon Principle.

In a general way, one can see that for a solute having an excited state much more polar than the ground state, increase of solvent polarity will stabilize the excited state more than the ground state, thereby lowering the energy of the transition and leading to a predicted red shift. Such is observed with the merocyanine dyes which contain the chromophore (2.33) and whose first excited state is approximated by dipolar structures such as (2.34).

† In such systems the differing lifetimes of singlet and triplet states may be a critical factor in determining behaviour.

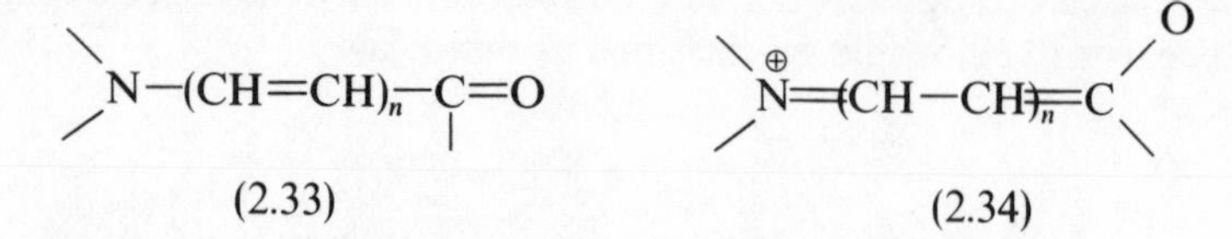

Conversely, the quaternary pyridinium iodide (2.35), whose absorption spectrum is remarkably solvent dependent, shows a marked *blue* shift, which has been rationalized[39] by assuming that its long wavelength absorption is due to a charge–transfer transition involving the transfer of charge from the iodide ion to the aromatic ring. Here, the excited state, with a large contribution from structure (2.36), is considerably *less* polar than the ground state. The magnitude of the solvent shifts for this compound has been proposed as an empirical scale for classifying solvent polarity.

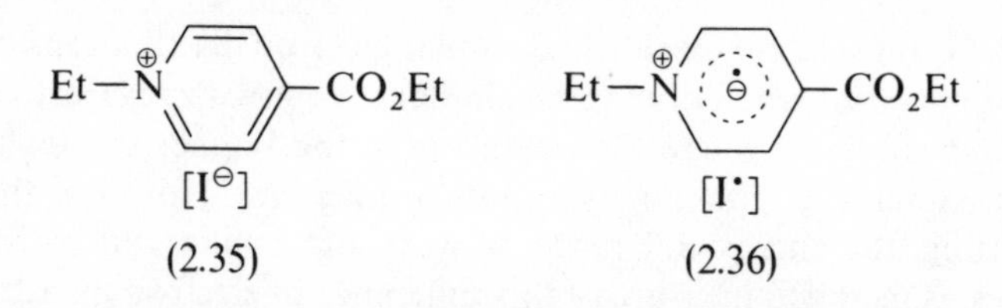

The blue shift associated with ($n \rightarrow \pi^*$) transitions, first noted by Kasha,[40] has now been extensively documented. It is so general a phenomenon that it has acquired a diagnostic importance. The observation of a low-intensity transition showing a shift of λ_{max} to shorter wavelengths on passing from hydrocarbons or carbon tetrachloride to alcohols as solvents is strong presumptive evidence of an $n \rightarrow \pi^*$ transition (see Table 2.4 for data for acetone). The origin of this effect

Table 2.4. Solvent shifts: λ_{max} of Me_2CO in nanometres

Solvent	H_2O	MeOH	EtOH	$CHCl_3$	CCl_4	C_nH_{2n+2}
λ_{max}	264	270	272	276	280	280

is not completely clear. With respect to aliphatic ketones, there is a small reduction in dipole moment on excitation, and this would be expected to lead to a small blue shift. Other factors, however, and notably the Franck–Condon Principle, probably play an important role. Around a polar solute molecule in its ground state, the solvent molecules are oriented so as to minimize the total free energy of the system. Electronic excitation is so rapid a process that the solvent cage surrounding the excited species is identical with that which surrounded the ground state species, and it will not in general have the configuration appropriate to the different charge distribution in the excited state. Hence the solvation energy of the solute in its excited state may well be less than that of the solute in its ground state. Since this effect will increase with increasing polarity of the solvent molecules, a blue shift will result.[41] Differences in hydrogen bonding between solvent and ground state or excited state will also contribute to the blue shift in hydroxylic solvents.[42]

Polar solvents stabilize (π, π^*) states and destabilize (n, π^*) states with respect to the situation in hydrocarbon solvents. Thus, if molecular (π, π^*) and (n, π^*) states have comparable energies, changing the solvent may invert the energetic ordering of the levels (Figure 2.47) with dramatic results.

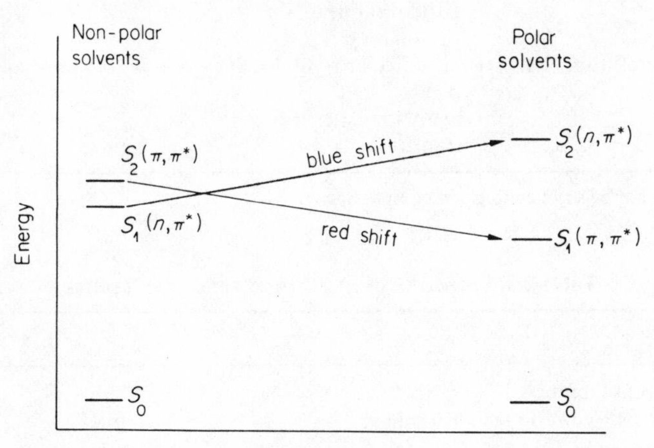

Figure 2.47. The effect of solvent polarity on the ordering of (n, π^*) and (π, π^*) states

For example, 2-naphthaldehyde and anthracene-9-carbaldehyde both fluoresce in ethanol but are non-fluorescent in heptane. The explanation probably resides in the fact that whereas $^1(\pi, \pi^*)$ states are frequently fluorescent, fluorescence from $^1(n, \pi^*)$ states is usually weak or undetectable (for reasons given in Chapter 3, p. 97). Hence, in heptane rapid internal conversion causes population of the lowest energy $^1(n, \pi^*)$ (non-fluorescent) state, whereas in ethanol it is the fluorescent $^1(\pi, \pi^*)$ state which becomes populated because it is the state of lowest energy.

2.7 IDENTIFICATION OF (n, π^*) AND (π, π^*) STATES

Since photochemical reactions are predominantly a function of the lowest excited singlet or triplet states, it is necessary to be able to identify the nature of this state. For this purpose appeal is usually made to spectroscopic data. Tables 2.5 and 2.6 set out the differences frequently encountered between the (n, π^*) and (π, π^*) states of organic molecules. It must be emphasized that many exceptions are known to the generalizations in the Tables, and for the identification of a state to be secure, it is essential that it conform to several of the criteria mentioned.

Table 2.5. Properties of singlet (n, π^*) and (π, π^*) states

	$^1(n, \pi^*)$	$^1(\pi, \pi^*)$
Intensity of absorption	*Weak*. Absorption band absent in hydrocarbon analogues	*Strong* (unless symmetry forbidden)
Solvent effect	Blue shift in polar or hydroxylic solvents. Band disappears on protonation	Red shift in polar solvents
Polarization of transition	Perpendicular to molecular plane	Parallel to molecular plane
Energy	Usually lowest energy[a] transition:	—

[a] This may not be so in conjugated compounds.

Table 2.6. Properties of triplet (n, π^*) and (π, π^*) states

	$^3(n, \pi^*)$	$^3(\pi, \pi^*)$
Phosphorescent lifetime	$<10^{-1}$–10^{-2} s	>1 s
Phosphorescence vibrational structure	Prominent	Variable
Singlet–triplet splitting	<36 kJ (8 kcal) mol^{-1} for C=O; <60 kJ (14 kcal) mole^{-1} for azines	>60 kJ (14 kcal) mol^{-1}
Intensity of $S_0 \rightarrow T$ transition	$f \sim 10^{-5}$–10^{-7}	$f \sim 10^{-11}$ (in hydrocarbons) 10^{-9}–10^{-6} (in haloaromatics)
External heavy atom effect on $S_0 \rightarrow T$ transition intensity	Little effect	Increases intensity
E.s.r.	No e.s.r. spectrum	Gives e.s.r. spectrum

REFERENCES

1. W. Kauzmann, *Quantum Chemistry*, Academic Press, New York (1957), ch. 16.
2. J. B. Birks and D. J. Dyson, *Proc. Roy. Soc.*, **A275**, 135 (1963); S. J. Strickler and R. A. Berg, *J. Chem. Phys.*, **37**, 814 (1962).
3. S. P. McGlynn, T. Azumi and M. Kinoshita, *Molecular Spectroscopy of the Triplet State*, Prentice-Hall, New Jersey (1969), p. 199.
4. R. S. Becker, *Theory and Interpretation of Fluorescence and Phosphorescence*, Wiley-Interscience, London, (1969), p. 82.
5. Ref. 4, p. 218.
6. D. F. Evans, *J. Chem. Soc.*, 1351 (1957).
7. J. B. Birks, *Photophysics of Aromatic Molecules*, Wiley-Interscience, London (1970), p. 495.

8. J. R. Platt, *J. Chem. Phys.*, **17**, 484 (1949).
9. W. C. Price and W. C. Tutte, *Proc. Roy. Soc.*, **A174**, 207 (1940).
10. J. E. Lennard-Jones, *Proc. Roy. Soc.*, **A158**, 280 (1937).
11. H. L. McMurry, *J. Chem. Phys.*, **9**, 231 (1941).
12. J. N. Murrell, *The Theory of the Electronic Spectra of Organic Molecules*, Methuen, London (1963), p. 161.
13. R. S. Mulliken, *Accounts Chem. Res.*, **9**, 7 (1976); C. Sandorfy in I. G. Csizmadia (ed.), *Applications of Molecular Orbital Theory in Organic Chemistry*, Elsevier, Amsterdam (1977), p. 384.
14. See ref. 7, p. 489 for relevant references.
15. G. Briegleb, *Elektronen-Donator-Acceptor-Komplexe*, Springer, Berlin (1961).
16. D. F. Evans, *J. Chem. Phys.*, **23**, 1424 (1955); *ibid.*, 1426.
17. H. A. Benesi and J. H. Hildebrand, *J. Amer. Chem. Soc.*, **71**, 2703 (1949).
18. J. N. Murrell, *J. Amer. Chem. Soc.*, **81**, 5037 (1959).
19. E. J. Bowen, *Chemical Aspects of Light*, 2nd edition, Clarendon Press, Oxford (1946).
20. For a review of chemiluminescence, see F. McCapra, *Quart. Rev.*, **20**, 485 (1966); W. R. Ware (ed.), *Creation and Detection of the Excited State*, Volume 3, Dekker, New York (1974), chapter 1.
21. Reviewed by A. Zweig, *Adv. Photochem.*, **6**, 425 (1968).
22. A. U. Khan and M. Kasha, *J. Amer. Chem. Soc.*, **88**, 1574 (1966).
23. F. McCapra, *Chem. Commun.*, 154 (1968).
24. O. Svelto, *Principles of Lasers*, Plenum, New York (1976).
25. (a) J. K. Burdett and M. Poliakoff, *Chem. Soc. Rev.*, **3**, 293 (1974); (b) F. P. Schäfer, *Angew. Chem. Intern. Ed.*, **9**, 9 (1970); (c) B. B. Snavely in J. B. Birks (ed.), *Organic Molecular Photophysics*, Volume 1, Wiley, London (1973), p. 239.
26. S. Kimel and S. Speiser, *Chem. Rev.*, **77**, 437 (1977).
27. D. A. Long, *Chem. in Brit.*, **7**, 108 (1971).
28. (a) V. S. Letokhov, *Science*, **180**, 451 (1973); (b) C. B. Moore, *Accounts Chem. Res.*, **6**, 323 (1973); (c) A. C. Vikis, *Chem. Phys. Lett.*, **53**, 565 (1978).
29. R. M. Hochstrasser, H.-N. Sung and J. W. Wessel, *J. Amer. Chem. Soc.*, **95**, 8179 (1973).
30. For references and discussion, see J. N. Murrell, *The Theory of the Electronic Spectra of Organic Molecules*, Methuen, London (1963).
31. A. J. Merer and R. S. Mulliken, *Chem. Rev.*, **69**, 642 (1969).
32. J. C. D. Brand and D. G. Williamson, *Adv. Phys. Org. Chem.*, **1**, 401 (1963); ref. 3, p. 166.
33. Reviewed by E. Vander Donckt, *Progr. Reaction Kinetics*, **5**, 273 (1970); ref. 4, p. 239.
34. F. Wilkinson in G. M. Burnett and A. M. North (ed.), *Transfer and Storage of Energy by Organic Molecules*, Wiley-Interscience, London (1969), p. 114.
35. E. Lippert, *Z. Electrochem.*, **61**, 962 (1957).
36. J. Czekalla, *Z. Electrochem.*, **64**, 1221 (1960).
37. L. Salem and C. Rowland, *Angew. Chem. Inter. Ed.*, **11**, 92 (1972).
38. See ref. 7, p. 115 for relevant references; P. Haberfield, M. Lux and D. Rosen, *J. Amer. Chem. Soc.*, **99**, 6828 (1977).
39. E. M. Kosower, *J. Amer. Chem. Soc.*, **80**, 3253 (1958).
40. M. Kasha, *Discuss. Faraday Soc.*, **9**, 14 (1950).
41. H. McConnell, *J. Chem. Phys.*, **20**, 700 (1952); G. C. Pimentel, *J. Amer. Chem. Soc.*, **79**, 3323 (1957).
42. N. S. Bayliss and E. G. McRae, *J. Phys. Chem.*, **58**, 1002 (1954); G. J. Brealey and M. Kasha, *J. Amer. Chem. Soc.*, **77**, 4462 (1955).

Chapter 3

Excited States: Time-dependent Phenomena

3.1 INTRODUCTION

In the previous chapter the factors relating to the production of excited states and their static properties were discussed. We now wish to analyse the time-dependent evolution of excited species into other entities, which may be alternative states of the same species (photophysical processes) or of other molecules (photochemical processes). This distinction between photophysics and photochemistry may be more apparent than real, but it will be retained for pedagogical reasons.

Excited states are short-lived, for they are compelled to lose their electronic energy within a short period of time. Even if no competing process intervenes, excited molecules must collapse to their ground state by the emission of radiation. Competing physical or chemical processes can give rise to a new excited state and hence delay momentarily the total loss of electronic energy, but the ultimate and rapid fate of all excited states is collapse to a ground state system. It is with some aspects of this area that the present chapter is concerned.

3.1.1 Dissipative Pathways

The processes responsible for the dissipation of the excess energy of an excited species may be differentiated and classified as in Figure 3.1.

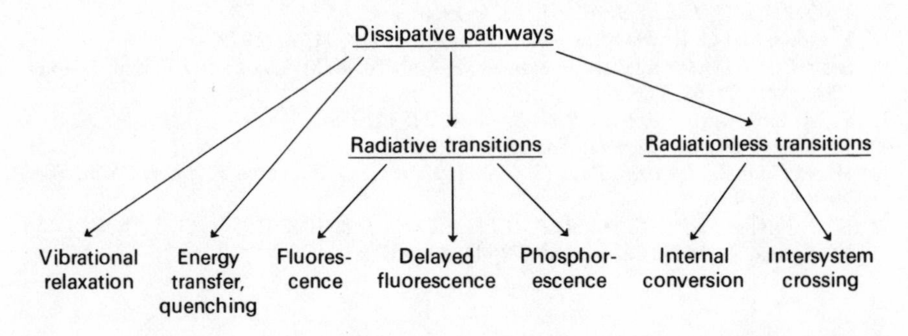

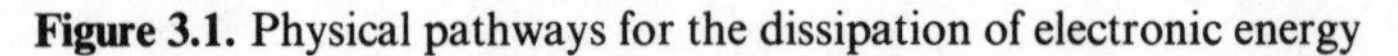

Figure 3.1. Physical pathways for the dissipation of electronic energy

It should be realized that these processes can compete with each other for the deactivation of an excited state, and the relative magnitude of the rate constants determines the contribution made by a particular pathway. To take a simple example, if an excited species can either fluoresce (rate constant k_f) or undergo some non-radiative deactivation (rate constant k_D), then if $k_D \gg k_f$ the population of excited species will be depleted predominantly by the radiationless route so that fluorescence will be weak and possibly undetectable. Conversely, if k_D and k_f are of comparable magnitude, strong fluorescence will be observed.

Vibrational Relaxation

Unless formed by a $0 \longrightarrow 0$ transition,† an excited species finds itself endowed at the moment of its creation with excess vibrational (and rotational) energy in addition to the electronic energy. The rate constant for the emission of infrared photons is so small due to the operation of the 'ν^3 Law' (see Chapter 2, p. 13) that loss of vibrational energy (called *vibrational relaxation* or *vibrational cascade*) is largely dependent upon collisions, as a result of which vibrational energy is converted into kinetic energy distributed between the partners in the collision. Consequently, in low pressure vapours where the time interval between collisions is greater than the lifetimes associated with radiative processes, emission is observed to come from the vibrational level populated by the absorption act.‡ As the pressure increases, collisions occur within the lifetime of the excited species and lead to a progressive loss of vibrational quanta, which is manifested by emission at longer wavelengths. At sufficiently high pressures, or in solution where the collision rate is of the order of 10^{13} s^{-1}, total vibrational relaxation is the rule and emission occurs almost exclusively from the $v' = 0$ level. There is, of course, a Boltzmann distribution over the vibrational levels, but at room temperature it is the zero-point level which is populated predominantly.

3.1.2 Radiative Transitions

In radiative transitions, represented by straight arrows on a Jablonski diagram (Figure 3.2), an excited species passes from a higher excited state to a lower one with the emission of a photon. Three processes may be distinguished:

(i) *Fluorescence* is caused by a radiative transition between states of the same multiplicity, and it is a rapid process ($k_f \sim 10^6$–10^9 s^{-1}). For the polyatomic molecules encountered in organic chemistry, the transition is usually $S_1 \longrightarrow S_0$, although $S_2 \longrightarrow S_0$ fluorescence is occasionally observed (e.g. with azulene or some thiocarbonyl compounds) and may often be hidden under the more

† A transition between the zero-point vibrational levels ($v = 0$) of the ground and excited states is referred to as a $0 \longrightarrow 0$ transition.

‡ This statement is not strictly correct. There is now good evidence that some systems with very sharp absorption bands, when excited under isolated molecule conditions, emit a structureless fluorescence. This phenomenon seems to mean that the initially populated vibrational level rapidly and non-radiatively redistributes its energy in a unimolecular fashion into a dense manifold of other isoenergetic levels, which then emit. Consideration of non-radiative transitions is deferred to later in this chapter.

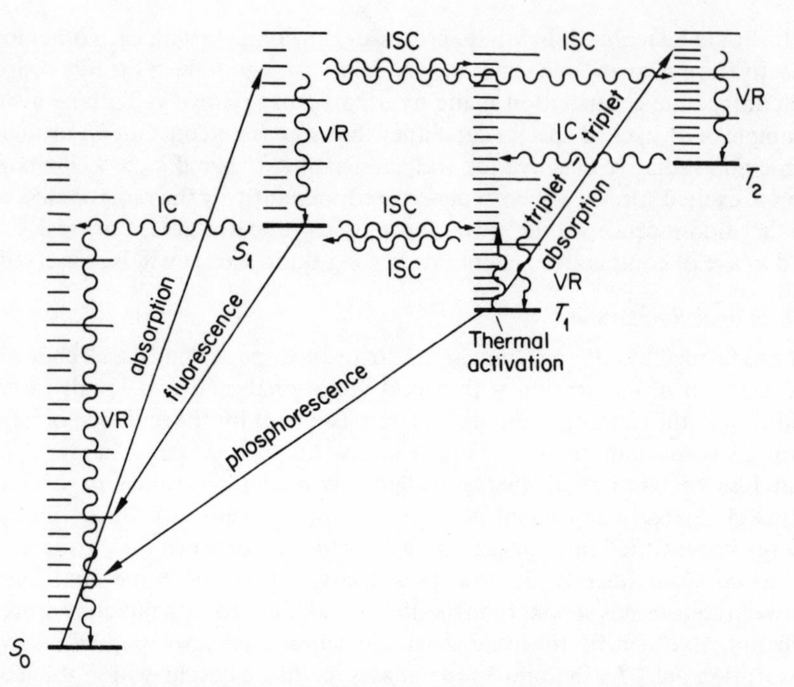

Figure 3.2. Jablonski diagram showing some of the radiative and non-radiative processes available to molecules (VR = vibrational, relaxation; IC = internal conversion; ISC = intersystem crossing)

intense $S_1 \longrightarrow S_0$ emission. A very weak $S_n \longrightarrow S_m$ emission has recently been detected for a number of other molecules, as has $T_n \longrightarrow T_m$ fluorescence. In sharp contrast, strong $S_n \longrightarrow S_m$ fluorescence is observed in diatomics.

(ii) *Phosphorescence* is the result of a transition between states of different multiplicity, typically $T_1 \longrightarrow S_0$; $T_n \longrightarrow S_0$ is very rare. The process, being spin forbidden, has a much smaller rate constant ($k_p \sim 10^{-2}$–10^4 s^{-1}) than that for fluorescence.

(iii) *Delayed Fluorescence* differs from ordinary fluorescence in that the measured rate of decay of emission is less than that expected from the transition giving rise to the emission.

3.1.3 Radiationless (or non-radiative) Transitions

Radiationless transitions occur between *isoenergetic* (or degenerate) vibrational–rotational levels of different electronic states. Since there is no change in the total energy of the system, no photon is emitted, and the process is represented by a horizontal line on a Jablonski diagram. Wavy arrows are used (e.g., $S_1 \rightsquigarrow T_1$) to distinguish radiationless transitions from radiative ones. If the electronic states participating in the radiationless transition are different states of the same molecule, then the transition is a *photophysical process* (e.g., internal conversion or intersystem crossing). However, a radiationless transition taking place between the excited state of one molecule and a state (usually the ground state) of another molecule gives rise to a *photochemical* transformation. Seen against this background, photochemistry is a facet of the general study of radiationless transitions (this point is discussed later).

Internal Conversion is a radiationless transition between isoenergetic states of the same multiplicity. Such transitions between upper states (e.g., $S_m \rightsquigarrow S_n$ or $T_m \rightsquigarrow T_n$) are extremely rapid, accounting for the negligible emission from upper states. Internal conversion from the first excited singlet state ($S_1 \rightsquigarrow S_0$) is so much slower that fluorescence can compete.

Intersystem Crossing is a radiationless transition between states of different multiplicity. The radiationless deactivation of the lowest triplet ($T_1 \rightsquigarrow S_0$) is a process in competition with normal phosphorescence. The intersystem crossing $S_1 \rightsquigarrow T_1$ or $S_1 \rightsquigarrow T_n$, which is competitive with (and reduces the quantum yield of) fluorescence, is the process by which the triplet manifolds are normally populated. $S_n \rightsquigarrow T_n$ has been observed but is rare because it has to compete with extremely fast internal conversion to S_1. The transition $T_1 \rightsquigarrow S_1$ requires thermal activation of T_1 to a vibrational level isoenergetic with S_1—it is the basis of one of the mechanisms leading to delayed fluorescence.

A Jablonski diagram (Figure 3.2) is frequently used to display the various exciting and dissipative pathways (compare the simplified diagram introduced in Chapter 1, Figure 1.6).

3.1.4 Kinetics, Quantum Yields and Lifetimes

Since kinetic analysis provides a powerful tool for unravelling the complexities of many photochemical phenomena, certain quantitative relations involving experimental quantities such as quantum yield and lifetime will now be established to show how the key rate constants may be extracted from such data.

First, consider a situation such as that depicted in Figure 3.3 where an excited species A* is subject to several first order or pseudo-first order deactivating processes

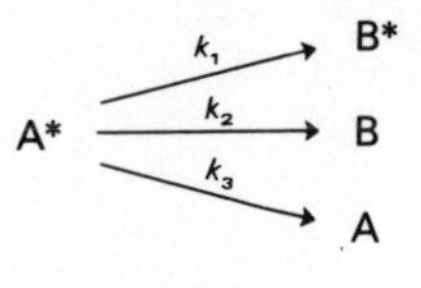

Figure 3.3

It follows that:

$$\frac{d[A^*]}{dt} = -[A^*](k_1 + k_2 + k_3) = -\sum k[A^*]$$

whence

$$[A^*] = [A_0^*]\, e^{-\Sigma kt}$$

The concentration of A* falls exponentially with a rate constant given by $\sum k$, and if the lifetime (τ) of A* is defined as the time taken for [A*] to fall to 1/e of its initial value, then

$$\tau = \frac{1}{\sum k} \tag{3.1}$$

τ, which is the actual measured lifetime of A*, should be clearly distinguished from the radiative lifetime τ_0 (see Chapter 2, section 2.1.2), which would be the lifetime if decay occurred exclusively by emission.

Now consider the general kinetic scheme set forth in Figure 3.4, where I is the rate of absorption of photons and k_f and k_p are the rate constants for fluorescence and phosphorescence.

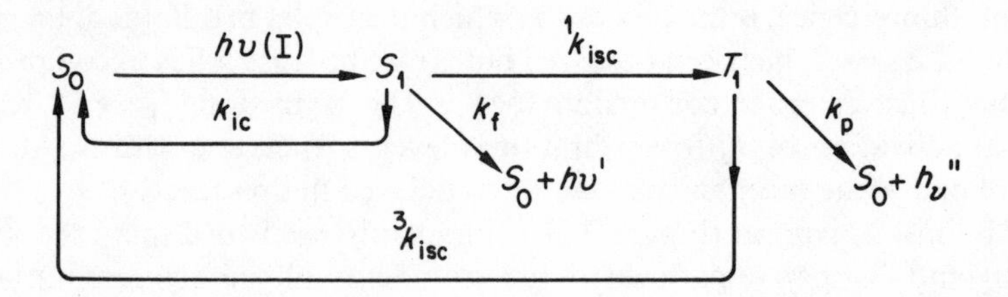

Figure 3.4. A general kinetic scheme for photophysical processes

The steady state approximation is applicable to excited states, and it follows that the rates of formation and destruction of S_1 are equal.

$$\therefore \quad I = [S_1]\sum {}^1k \quad (\text{where} \sum {}^1k = k_{ic} + k_f + {}^1k_{isc}) \tag{3.2}$$

The quantum yield of fluorescence is given by

$$\phi_f = \frac{\text{rate of emission by } S_1}{\text{rate of absorption of photons by } S_0} = \frac{k_f[S_1]}{I} = \frac{k_f}{\sum {}^1k} \tag{3.3}$$

and

$$\tau_f = \frac{1}{\sum {}^1k} \tag{3.4}$$

Hence

$$k_f = \frac{\phi_f}{\tau_f} \tag{3.5}$$

This permits calculation of k_f from experimental data. Since $k_f = A_{ul}$, the Einstein coefficient of spontaneous emission, it may also be obtained from absorption spectra by measuring the oscillator strength f and calculating B_{ul} and hence A_{ul} (see Chapter 2, p. 13). k_f is also related to τ_0 by the equation

$$\tau_0 = \frac{1}{k_f} \tag{3.6}$$

Therefore

$$\phi_f = \tau_f/\tau_0 \tag{3.7}$$

The steady state approximation applied to T_1 gives

$$^1k_{isc}[S_1] = \sum {}^3k[T_1] \quad (\text{where} \sum {}^3k = k_p + {}^3k_{isc}) \tag{3.8}$$

The quantum yield of phosphorescence is:

$$\phi_p = \frac{k_p[T_1]}{I} = \frac{k_p}{I} \cdot \frac{^1k_{isc}[S_1]}{\sum {}^3k} \quad (\text{from 3.8}) \tag{3.9}$$

$$= \frac{k_p}{\sum {}^3k} \cdot \frac{^1k_{isc}}{\sum {}^1k} \quad (\text{from 3.2}) \tag{3.10}$$

Hence

$$\phi_p = \theta_p \,.\, \theta_{isc} \tag{3.11}$$

where θ_p and θ_{isc} are the *quantum efficiencies* of phosphorescence and intersystem crossing respectively. Quantum efficiency (to be distinguished from quantum yield) is the ratio of the rate of a process involving an excited state to the rate of production of that state. Thus

$$\theta_p = \frac{k_p[T_1]}{\text{rate of production of } T_1} = \frac{k_p[T_1]}{\sum {}^3k[T_1]} \tag{3.12}$$

because the rates of production and destruction of T_1 are equal (steady state approximation). Hence

$$\theta_p = \frac{k_p}{\sum {}^3k} \tag{3.13}$$

Similarly,

$$\theta_{isc} = \frac{^1k_{isc}[S_1]}{I} = \frac{^1k_{isc}[S_1]}{\sum {}^1k[S_1]}$$

$$= \frac{[T_1]\sum {}^3k}{[S_1]\sum {}^1k} = \frac{\sum {}^3k[T_1]}{I} \quad (\text{from 3.2 and 3.8}) \tag{3.14}$$

$$\therefore \quad \theta_{isc} = \frac{\text{rate of production of } T_1}{\text{rate of absorption of photons}} = \frac{\text{rate of destruction of } T_1}{\text{rate of absorption of photons}}$$

Hence

$$\theta_{isc} = \phi_T \quad (\text{the quantum yield for formation of triplets})$$

Since

$$\phi_T = \theta_{isc} = \frac{^1k_{isc}}{\sum {}^1k}$$

and

$$\tau_f = \frac{1}{\sum {}^1k}$$

it follows that

$$^1k_{isc} = \frac{\phi_T}{\tau_f} \tag{3.15}$$

Thus can $^1k_{isc}$ be calculated. k_p can, in principle, be obtained knowing θ_p and τ_p, for from (3.13)

$$\theta_p = k_p \tau_p, \quad \text{where } \tau_p = \frac{1}{\sum {}^3k}$$

Since θ_p cannot be measured directly, use is made of ϕ_p instead, because from (3.9)

$$\frac{\phi_p}{k_p} = \frac{[T_1]}{I} = \frac{\sum {}^3k[T_1]}{\sum {}^3k \,.\, I} = \tau_p \,.\, \phi_T$$

Hence

$$k_p = \frac{\phi_p}{\phi_T \,.\, \tau_p} \tag{3.16}$$

ϕ_T, the quantum yield of triplet production, is obtained from triplet counting techniques (see Chapter 5, p. 156).

3.2 RADIATIVE TRANSITIONS[1,2]

3.2.1 Methods

Since upper states do not generally luminesce, the methods just described cannot be employed to obtain k_{ic} among upper states, and consequently there is little information about this rate constant. An order of magnitude estimate may be derived from an argument due to Kasha[3] that since fluorescence from upper singlets would have been detected were it 10^{-4} times as intense as the ordinary fluorescence, then k_{ic} must be $\geqslant 10^4\, k_f$ (where k_f is rate constant for the unobserved fluorescence). If k_f is estimated from the ultraviolet absorption spectrum to be $> 10^8\ s^{-1}$, $k_{ic} > 10^{12}\ s^{-1}$ and $\tau < 10^{-12}$ s. One way of obtaining the lifetime of upper states is to examine the width of absorption lines. Because of the Uncertainty Principle a short-lived state would have a relatively broad line. Currently accepted values of k_{ic} are $\sim 10^{12}\ s^{-1}$.

Fluorescence Spectra

A spectrofluorimeter (Figure 3.5) is the instrument of choice for recording fluorescence spectra. A beam of monochromatic light excites the specimen in the cell, and the emission is observed and analysed at right angles to the incident beam. The output is the emission spectrum plotted by the XY recorder.

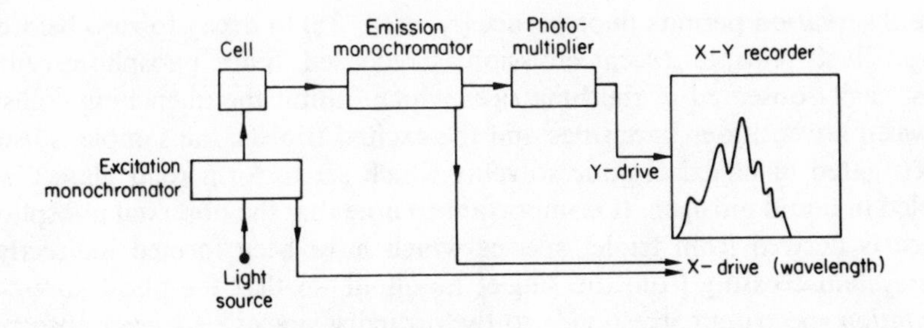

Figure 3.5. Schematic diagram of a spectrofluorimeter

Since with few exceptions (notably biacetyl) organic molecules do not phosphoresce in ordinary solvents at room temperature, the system outlined will record just the fluorescence spectrum, and since either the excitation or the emission monochromator may be coupled to the XY recorder it follows that two sorts of spectra may be obtained:

(i) Set the excitation monochromator to a particular wavelength absorbed by the sample and scan the emitted light with the emission monochromator. This affords the *fluorescence emission spectrum.*

(ii) Set the emission monochromator to a particular wavelength in the fluorescent output and scan the exciting wavelengths with the exciting monochromator. This gives rise to the *fluorescence excitation spectrum*, which normally closely resembles† the absorption spectrum in sufficiently dilute solutions (see Figure 3.8).

Fluorescence excitation spectroscopy can be used to identify and quantitatively to estimate fluorescent molecules.[4] It supplements absorption spectroscopy, but with the difference that fluorimetric analysis is often orders of magnitude more sensitive. This increase in sensitivity is primarily due to the fact that the fluorescence signal can be observed, amplified and recorded directly, whereas with absorption spectroscopy what is measured is the signal due to the *difference* between the incident and transmitted light intensities. Fluorimetry provides a powerful tool for the analysis of mixtures. Often only one component is fluorescent, and when more than one component is fluorescent it may well be possible selectively to excite just one of the components by adjusting the excitation wavelength.

Phosphorescence Spectra

These are recorded with spectrophosphorimeters, which differ from spectrofluorimeters only in the incorporation of a mechanical or optical shutter which repetitively chops the exciting and emitted light beams in such a manner that excitation occurs when the detector is cut off and emission is not observed until a definite period after excitation has ceased. This delay between excitation

† For a precise correspondence to be obtained with a single-beam instrument (Figure 3.5), the recorded spectrum should be corrected for the variation with wavelength of the emission of the light source, of the transmission of the monochromators and of the response of the photomultiplier.

and observation permits fluorescence ($\tau_f < 10^{-6}$ s) to decay to zero before the longer-lived phosphorescent emission is recorded. Since phosphorescence is most easily observed in rigid matrices which inhibit the quenching collisions between adventitious impurities and the excited triplets, the sample is usually investigated in mixed organic solvents which set to form rigid glasses when cooled in liquid nitrogen. It is important to note that the observed phosphorescence is derived from triplet species which have been formed indirectly by intersystem crossing from the singlet manifold, so that the phosphorescence *excitation* spectrum corresponds to the ordinary singlet $\rightarrow$ singlet absorption spectrum.

The extreme sensitivity associated with emission spectroscopy makes it possible to estimate triplets produced *directly* via forbidden $S \rightarrow T$ transitions. This forms the basis of *phosphorescence excitation spectroscopy.*[5] An intense light source is used to excite the $S \rightarrow T$ transition, a heavy atom solvent (e.g., ethyl iodide) being used where appropriate to increase the intensity of absorption. The production of excited triplets is then monitored by observing their phosphorescence at a particular set wavelength. The spectrum obtained by plotting the intensity of phosphorescence against the wavelength of the exciting light (at constant intensity) is the $S \rightarrow T$ absorption spectrum, because the intensity of the phosphorescence is directly proportional to the concentration of triplets which is itself proportional to the extinction coefficient for the $S \rightarrow T$ transition at the particular exciting frequency used (Figure 3.6) This technique may be used even when the compound itself is non-phosphorescent by having present a phosphorescent molecule of lower triplet energy which is excited by energy transfer (see Chapter 4) from the non-phosphorescent triplets. Some representative emission spectra are given in Figures 3.6–3.8. Figure 3.7 illustrates the general point that phosphorescence occurs at longer wavelengths

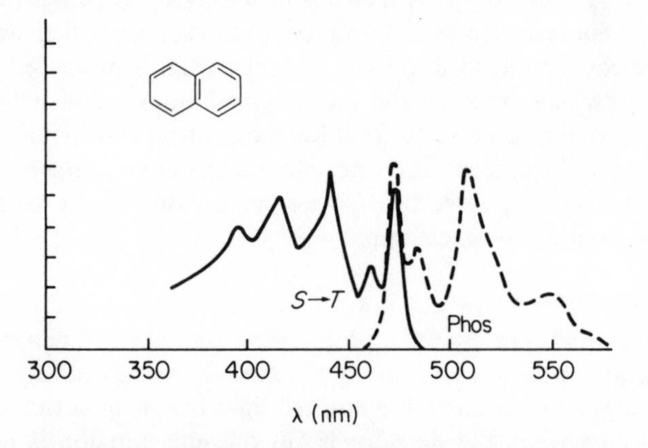

Figure 3.6. Phosphorescence spectrum and $S_0 \rightarrow T_1$ absorption spectrum from phosphorescence excitation of naphthalene (reproduced with permission from A. P. Marchetti and D. R. Kearns, *J. Amer. Chem. Soc.*, **89**, 768 (1967); copyright by the American Chemical Society)

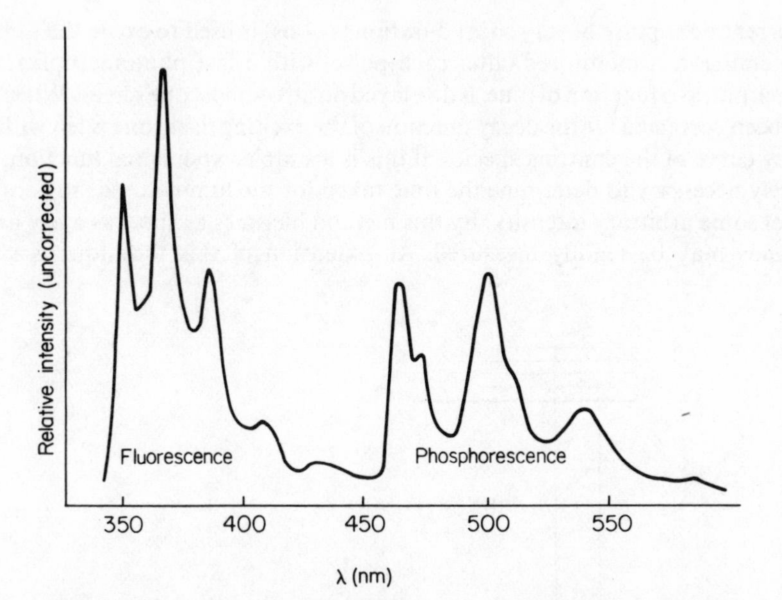

Figure 3.7. Total emission spectrum of phenanthrene (10^{-4} M) in ethanol at 77 K (from F. Wilkinson and A. R. Horrocks, in E. J. Bowen (ed.), *Luminescence in Chemistry* (1968), by permission of Van Nostrand Reinhold Co. Ltd.)

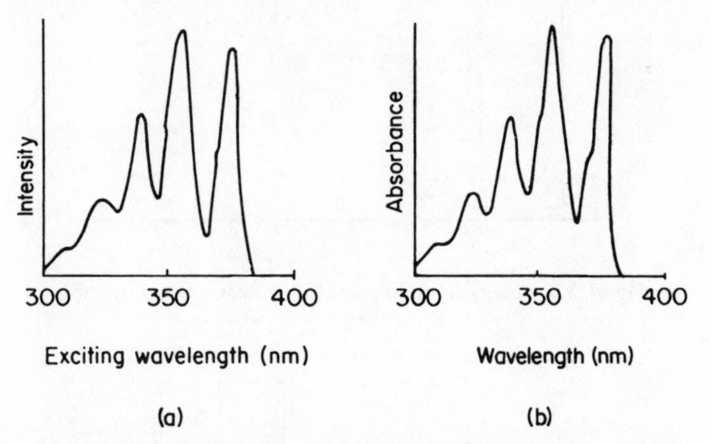

Figure 3.8. (a) Corrected fluorescence excitation spectrum, and (b) absorption spectrum, of anthracene

than fluorescence ($E_T < E_S$), and Figure 3.8 demonstrates that fluorescence excitation and absorption spectra are closely similar.

Lifetimes

Of the various methods available for measuring the actual lifetimes of excited states, the most direct is pulse fluorimetry or phosphorimetry, in which a recurrent light pulse of very short duration (~ 1 ns) is used to excite the sample. The emission is monitored (after each pulse) with a fast photomultiplier, and the output, as a function of time, is displayed on an oscilloscope screen. When this has been corrected for the decay function of the exciting flash, one is left with the decay curve of the emitting species. If this is a simple exponential function, it is merely necessary to determine the time taken for the luminescence to decay to 1/e of some arbitrary intensity. By this method lifetimes as short as a few nano-seconds may be readily measured. An extension of this technique is *single*

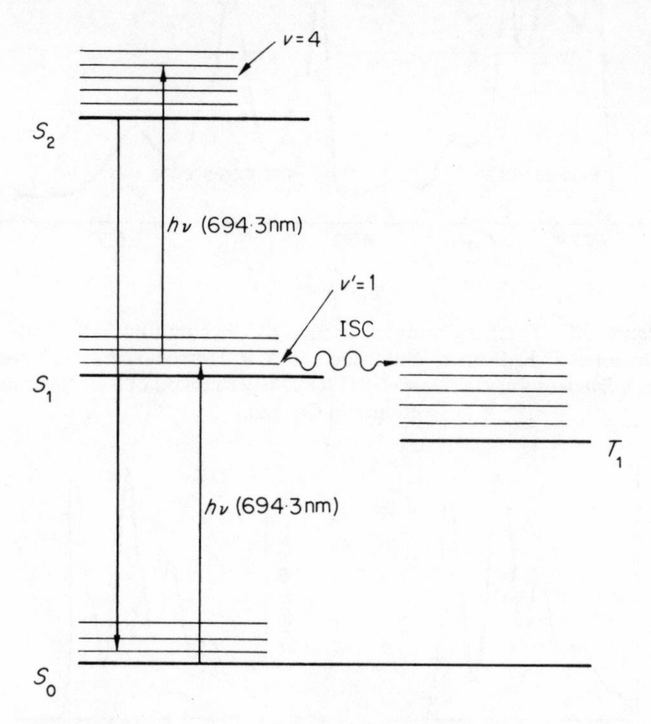

Figure 3.9. Energy level diagram for azulene excited by ruby laser

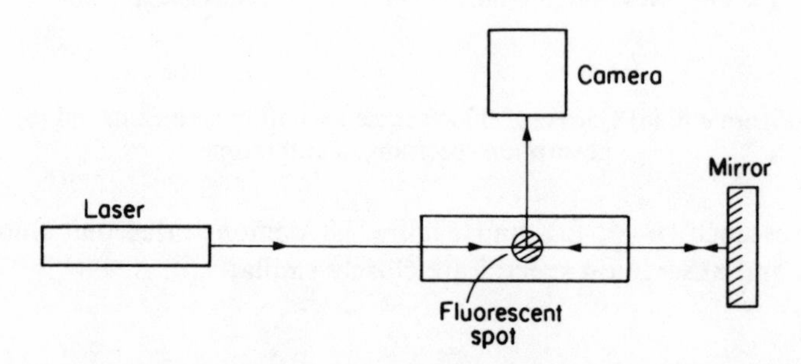

Figure 3.10. Schematic diagram of apparatus for measuring k_{isc} for azulene

photon counting[6] which, because it depends on counting individual photons, can be used on extremely weakly luminescent substances. It permits measurement of lifetimes in the range 10^{-6}–10^{-10} s. Still shorter lifetimes in the picosecond range depend for time measurement on the distance travelled by light. This technique, pioneered by Rentzepis, can be illustrated by the determination of k_{isc} for azulene.[7] Azulene was selected because (exceptionally) it fluoresces from S_2 and only negligibly from S_1. The output from a Q-switched mode-locked ruby laser consisting of a train of pulses ($\bar{\nu} = 14\,400$ cm^{-1}) of a few picoseconds bandwidth is passed through a solution of azulene. A given pulse, as it travels down the cell, excites azulene molecules in its path to the first vibrational level of the S_1 state (Figure 3.9). The excited azulene molecules start to decay immediately at a rate determined by the rate constant for radiationless processes ($S_1 \rightsquigarrow T_1$ and $S_1 \rightsquigarrow S_0$). Consequently the pulse leaves behind it a trail of decaying excited molecules. The pulse is reflected back along its original path by a mirror at the end of the cell (Figure 3.10). On its return journey it excites any remaining S_1 azulene molecules to the fluorescent S_2 state. A fluorescent spot is thus produced in the cell. The further the reflected pulse travels in its return journey, the less the probability of its encountering azulene molecules still remaining in the S_1 state. Hence the dimensions of the fluorescent spot, after correction for the pulse width, are related to the rate constants for radiationless processes. By these means it was estimated that the lifetime of the S_1 state of azulene was ~4 ps. The process primarily responsible for deactivating the S_1 state was shown to be the intersystem crossing $S_1 \rightsquigarrow T_1$. It was concluded that, in this case, $k_{isc} \sim 2{\cdot}5 \times 10^{11}$ s^{-1}.

Quantum Yields

Although in order to obtain the quantum yield of fluorescence one has in principle merely to measure the ratio of the number of photons emitted to those absorbed, in practice grave difficulties attend the determination arising from (i) the difference in the spatial distribution of the exciting and emitted light, (ii) the polychromatic character of the emitted light, and (iii) the variation of the sensitivity of the detector with wavelength.

The last two problems can be greatly simplified by directing the incident light and fluorescent emission successively onto a '*quantum counter*' (a solution of a substance such as Rhodamine B, which, within a certain range of wavelengths, converts all absorbed light at constant quantum yield into its own fluorescent emission). The detector then receives signals of constant spectral distribution from both incident and emitted light beams.

Given that the spectral response of the light detector is known, the most rapid and accurate way of determining emission efficiency is to measure the unknown quantum yield relative to that of some substance whose absolute emission quantum yield has already been accurately measured. It is then only necessary to determine, under identical conditions of cell geometry, incident light intensity and temperature, the fluorescence spectra of dilute solutions of the unknown and of the standard. The solutions should have the same optical density at the wavelength of the exciting light so that they both capture the same number of photons. The quantum yield of the unknown relative to that of the standard is the ratio of the integrated band areas under the two fluorescence spectra (plotted on a frequency scale) after they have been corrected for the detector response function. Multiplying by the known quantum yield of the standard then gives the absolute quantum yield of the unknown. The technique has been amply described.[8]

Polarization Spectra

In Chapter 2 it was seen that both absorption and emission are polarized along particular molecular axes determined by the symmetry properties of the participating orbitals and predictable by Group Theory. The method of obtaining polarization data and their applications are now considered.

With single crystals of known and appropriate structure, the orientation of the plane of polarization with respect to the crystal axes and hence the molecular axes may be determined. This gives the absolute polarization of absorption and emission. It is easier and often more useful to determine the *relative* polarization of absorption bands, and this information is given by polarization spectra in which the polarization is plotted against the wavelength of the exciting light. Such spectra may be obtained by inserting polarizing devices (P_1 and P_2) into the incident and emitted beams of light of a spectrofluorimeter or spectro phosphorimeter (Figure 3.11). With P_1 fixed, for each exciting frequency the

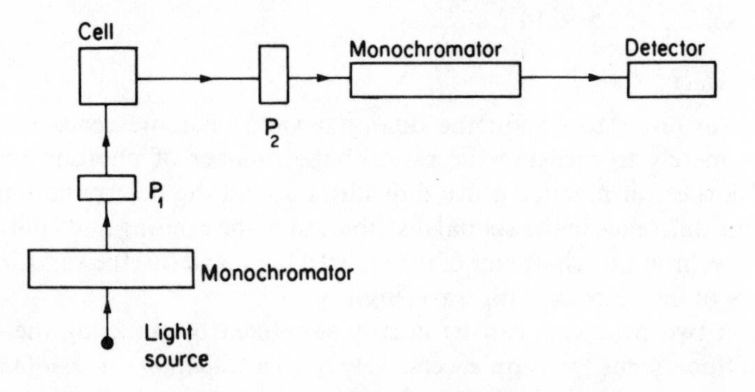

Figure 3.11. Schematic diagram of apparatus for determining polarization spectra

intensity of the emitted light is measured with the plane of polarization defined by P_2 parallel or perpendicular to that of P_1. This gives the *degree of polarization* (P) defined by

$$P = \frac{I_{\parallel} - I_{\perp}}{I_{\parallel} + I_{\perp}} \tag{3.17}$$

where $I_{\parallel}$ and $I_{\perp}$ are the intensities of the parallel and perpendicular components of the emitted light.

For randomly oriented molecules in a highly viscous solvent which inhibits rotation in the time interval between absorption and emission it can be shown that

$$P = \frac{3\cos^2\alpha - 1}{\cos^2\alpha + 3} \tag{3.18}$$

where α is the angle between the directions of polarization of absorption and emission of the substrate. Since α is commonly 0° or 90°, P can assume values ranging only between $+\frac{1}{2}$ and $-\frac{1}{3}$. In fact, such values are rarely achieved because of various depolarizing effects.

The absorption and fluorescence polarization spectra of phenol are shown in Figures 3.12 and 3.13. Notice that P changes sign at about 240 nm indicating that the $S_0 \rightarrow S_1$ and the $S_0 \rightarrow S_2$ transitions have different directions of polarization with respect to the molecular axes.

The apparently simple long wavelength absorption band of aniline ($\lambda_{max} \sim$ 283 nm) is revealed to be due to at least two different transitions by its polarization spectrum (Figure 3.14).

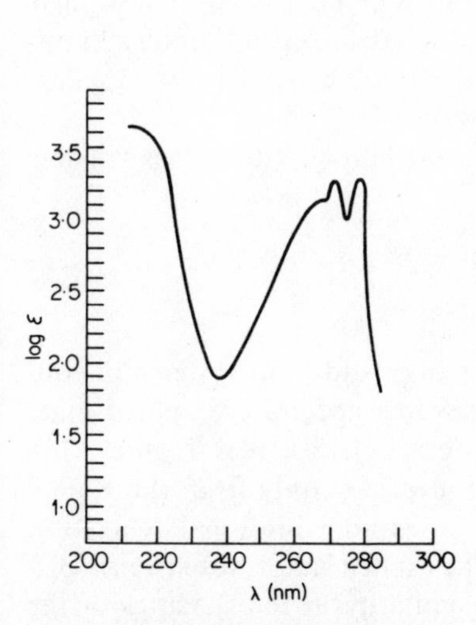

Figure 3.12. Ultraviolet absorption spectrum of phenol in cyclohexane (from R. A. Friedel and M. Orchin, *Ultraviolet Spectra of Aromatic Compounds*, (1951), John Wiley and Sons Ltd.)

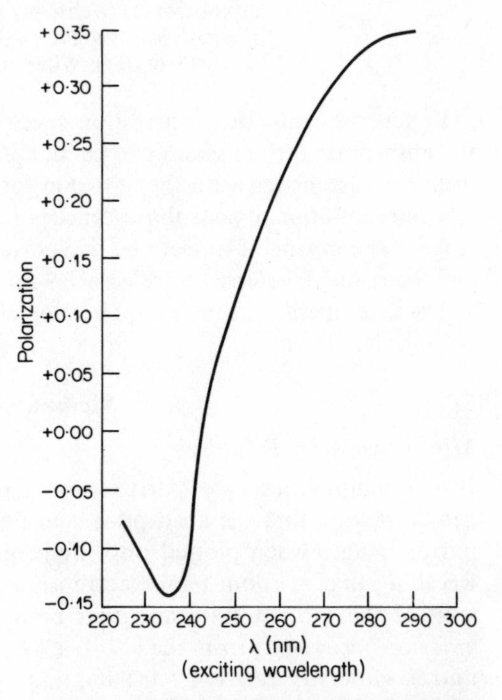

Figure 3.13. Fluorescence polarization spectrum of phenol at −70 °C in propylene glycol (from G. Weber, in D. M. Hercules (ed.), *Fluorescence and Phosphorescence Analysis*, (1966), John Wiley and Sons Ltd.)

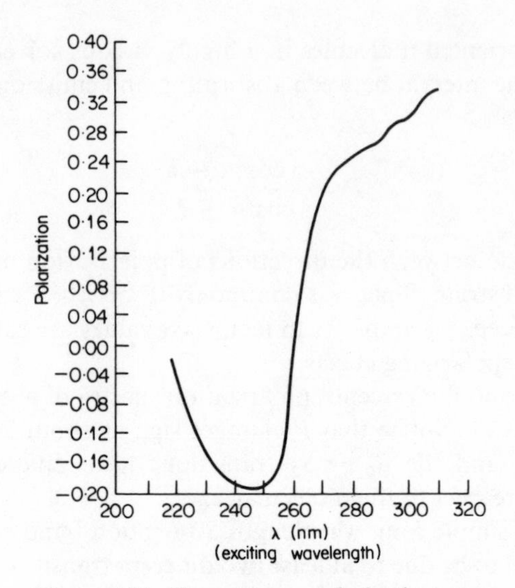

Figure 3.14. Fluorescence polarization spectrum of aniline at −70 °C in propylene glycol (from G. Weber, in D. M. Hercules (ed.), *Fluorescence and Phosphorescence Analysis*, (1966), John Wiley and Sons Ltd.)

In general, emission polarization spectroscopy gives a value for α, and, if the absorption (or emission) can be identified with a particular transition, the transition associated with the emission (or absorption) can often be assigned. The interpretation of phosphorescence polarization data depends upon a knowledge of the mixing of singlet and triplet states induced by spin–orbit coupling, and the reader is referred to reference 9 for details.

The data emerging from the application of the methods just described will now be considered.

3.2.2 Fluorescence

Mirror Symmetry Relation

It is commonly observed, particularly among large and rigid systems in condensed phases, that the absorption and fluorescence spectra are approximate mirror images when plotted on a frequency (energy) scale (e.g. Figure 3.15). Recalling that at room temperature absorption occurs only from the $v'' = 0$ level of the ground state and that because of rapid vibrational relaxation emission occurs only from the $v' = 0$ level of the excited state, the existence of a mirror symmetry relation must imply close similarity in the spacings of the vibrational levels in the ground and excited states. This in turn implies that only minor changes in geometry occur on excitation, which is to be expected with rigid and extended chromophores (see Chapter 2, p. 49). Furthermore, the Franck–Condon principle, which determines the shape of the absorption envelopes, must also apply to emission processes.

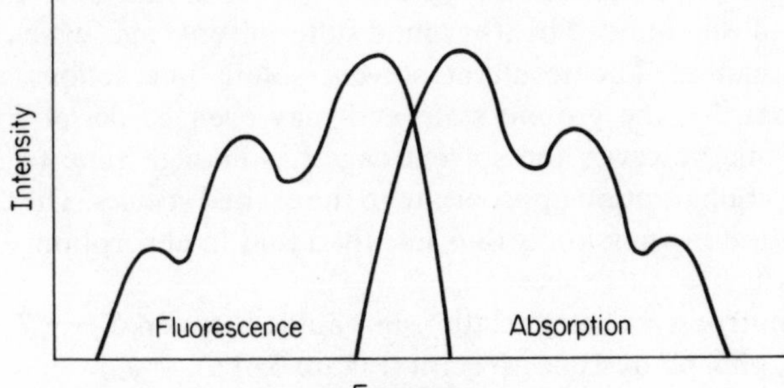

Figure 3.15. Typical mirror-symmetry relation between fluorescence and absorption spectra

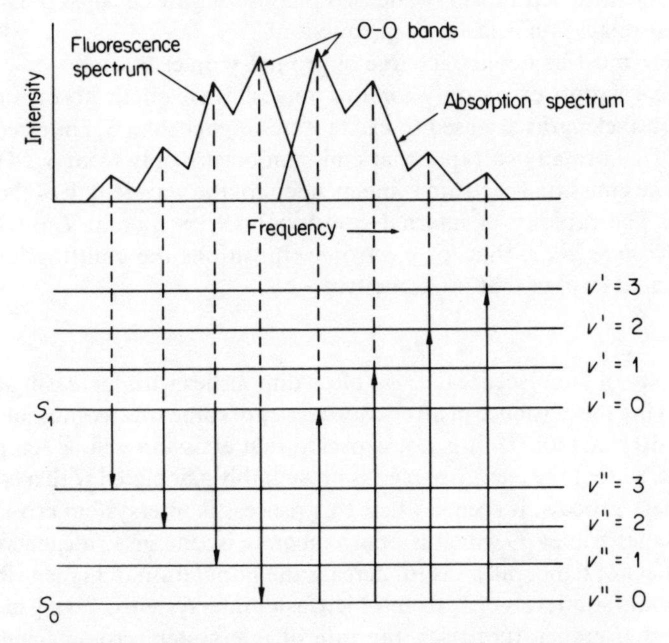

Figure 3.16. The origin of the mirror-symmetry relation

Figure 3.16 shows (i) that the vibrational spacings in the fluorescence spectrum should correspond to ground state energy levels (contrast absorption spectra), and (ii) that the two spectra should be symmetrically disposed about the 0—0 bands which indeed should be superimposed. This latter is actually observed in vapour phase spectra, but when spectra are recorded in solution there is a separation between the 0—0 bands of the fluorescence and absorption spectra, the magnitude of which is temperature and solvent dependent.

The separation arises from a Franck–Condon effect. The equilibrium solvent cages surrounding a molecule in its ground state and excited state will be different because of the changes in dipole moment and geometry occurring on excitation. Electronic excitation is so fast ($\sim 10^{-15}$ s) that after excitation the molecule is still surrounded by its ground state solvent cage, which has not had time to reorganize. The resultant solvent–solute interactions will be less stabilizing than for the ground state and may even be destabilizing. Before emission occurs, however, the solvent cage has enough time to relax to the lower energy configuration appropriate to the excited species. The energy of the $0 \rightarrow 0$ transition in emission is thus less than that in absorption, and the 0—0 bands separate.

A similar mirror symmetry relationship holds between $S_0 \rightarrow T_1$ absorption and $T_1 \rightarrow S_0$ phosphorescence spectra (Figure 3.6).

It should be recognized that the mirror symmetry relation will only be observed if certain conditions are fulfilled:

(i) It requires that the excited molecules emit from the $v' = 0$ level. Hence it is found only with spectra from condensed phases (or gases at such pressures that vibrational relaxation is faster than emission).

(ii) There must be no large charge of geometry on excitation.

(iii) The relation exists only for the longest wavelength absorption band. If shorter wavelengths are used to excite states higher than S_1, internal conversion to S_1 is normally so rapid that emission occurs only from S_1. Hence the fluorescence emission spectrum is independent of the wavelength of the exciting radiation. The rapidity of internal conversion ($S_n \rightsquigarrow S_1$ and $T_n \rightsquigarrow T_1$) is the basis of *Kasha's Rule*, that for electronic transitions the emitting level is the lowest excited level of that multiplicity.

Temperature

The intensity of fluorescence (i.e., ϕ_f) often diminishes with increasing temperature, implying the existence of an energy barrier of some sort (commonly 4–40 kJ mol^{-1}, 1–10 kcal mol^{-1}). It is not expected that emission will be temperature-dependent, so that the energy barrier is presumably associated with a competing radiationless process. It seems[10] that this process is intersystem crossing from S_1 to a higher triplet T_n which is approximately isoenergetic (degenerate) with S_1. The effect of temperature is to increase the population of higher vibrational and rotational sub-levels of S_1 from which faster intersystem crossing may occur. Hence as the temperature rises, the rate of intersystem crossing increases, a smaller proportion of S_1 molecules have an opportunity to fluoresce, and ϕ_f drops accordingly.

Quenching

If $(\phi_f + \phi_p) < 1$, then fewer photons are emitted than are absorbed and luminescence can be thought of as having been 'quenched'. Quenching may be partial or total, and it may be ascribed to internal factors (radiationless processes leading to the ground state) or to external factors (interactions of the excited state with other molecules). These quenching processes are extremely important, but, since they are analysed in detail later in this chapter and in Chapter 4, they will not be discussed further here.

3.2.3 Phosphorescence

Because of the long radiative lifetime of triplet states (typically 10^{-4}–10^{2} s) caused by the spin-forbidden nature of the emission process, they are particularly susceptible to quenching collisions with adventitious impurities. Thus phosphorescence (except that of biacetyl) is difficult to observe in the gas phase or in fluid solution. Although the use of highly purified solvents, particularly perfluorocarbons,[11] does permit useful observations of phosphorescence in fluid solution, most observations are made on solutions in rigid glasses in which the diffusion of quenchers is strongly inhibited. The commonly used phases are mixtures of organic solvents (e.g., ether, isopentane, ethanol) which, at 77 K, set to form glasses. For work at room temperature use is made of organic plastics or of melts in boric acid or other inorganic glasses.[12]

That the extremely feeble phosphorescence associated with solutions is not due to some effect on the production of the triplet species has been amply demonstrated by flash photolysis and energy transfer studies, which show that triplets can be formed in high yield in solutions whose phosphorescence is virtually non-existent.

Triplets and Phosphorescence

Although it has been implicitly assumed so far that phosphorescence is a property of the triplet state, it is only relatively recently that this identification has been established with certainty. A great deal of suggestive evidence has been collected, which has been excellently and critically reviewed by McGlynn *et al.*,[13] but the decisive observations concern the magnetic properties of the phosphorescent state.

(i) Triplet states, having two spin-parallel electrons, should be paramagnetic. Several authors have shown that fluorescein anion and other organic molecules when irradiated in a rigid glass showed both phosphorescence and paramagnetic susceptibility and that the lifetimes of both were identical (Figure 3.17).

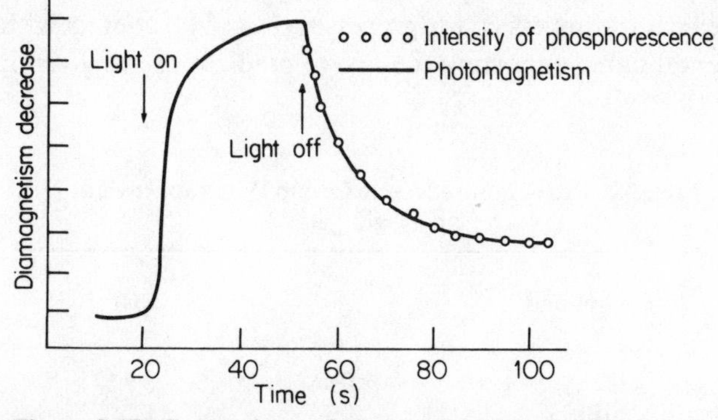

Figure 3.17. Comparison of photomagnetism and decay of phosphorescence of triphenylene in boric acid 'glass' at room temperature (from D. F. Evans, *Nature*, **176**, 777 (1955), reproduced by permission of Nature)

(ii) The triplet state has three degenerate components with the spin wave-functions given in Chapter 2 (p. 54) and magnetic quantum numbers $M_S = 1, 0, -1$. In a magnetic field the degeneracy is lifted, and transitions between the triplet sublevels become possible. These transitions can be detected by electron spin resonance (e.s.r.). The first successful application of e.s.r. was the detection of the paramagnetic resonance of the phosphorescent state of naphthalene. Subsequently it was shown that the intensities of phosphorescent emission and of the e.s.r. signal from irradiated aromatic ketones decayed at the same rate. The subject has been reviewed.[14]

Heavy Atom Effects

In Chapter 2 it was pointed out that the heavy atom effect markedly increased the rates of all singlet ↔ triplet processes, both radiative and non-radiative. The consequences with respect to molecular emission are set out in the following Table 3.1.

Table 3.1. Heavy atom effect on lifetimes and quantum yields of emission

Rate process accelerated by heavy atom effect	Expected consequences
(1) $S_0 \rightarrow T_n$	Enhanced $S_0 \rightarrow T_n$ absorption (Chapter 2)
(2) ISC ($S_1 \rightsquigarrow T_n$)	τ_f and ϕ_f decreased, ϕ_T increased
(3) ISC ($T_1 \rightsquigarrow S_0$)	τ_p decreased, ϕ_p decreased
(4) $T_1 \rightarrow S_0$	τ_p decreased, ϕ_p increased

Although it is clear that τ_p, τ_f and ϕ_f should all decrease, it is difficult to predict whether the intensity of phosphorescence (ϕ_p) will increase or decrease as a result of the heavy atom effect, for processes (2) and (4) conflict with process (3).

Experimental data supporting the above predictions are given in Tables 3.2 and 3.3.

Table 3.2. Heavy atom effects in Group IV tetraphenyls at 77 K in a rigid glass[15]

Compound	$\frac{\phi_p}{\phi_f}$	τ_p(s)
CPh_4	$\ll 0{\cdot}1$	2·9
$SiPh_4$	0·1	1·1
$GePh_4$	1	0·055
$SnPh_4$	10	0·003
$PbPh_4$	$\gg 10$	$<0{\cdot}001$

Table 3.3. Heavy atom effect in halonaphthalenes in a rigid glass at 77 K[16]

Substituted Naphthalene	ϕ_p	ϕ_f	$\frac{\phi_p}{\phi_f}$	τ_p(s)
Naphthalene	0·05	0·55	0·091	2·3
1-Fluoro-	0·056	0·84	0·067	1·5
1-Chloro-	0·30	0·058	5·2	0·29
1-Bromo-	0·27	0·0016	169	0·018
1-Iodo	0·38	<0·0005	>760	0·002

Deuteration

Perdeuteration markedly increases the phosphorescent lifetime τ_p for many aromatic hydrocarbons. For example, for naphthalene in durene at 77 K, $\tau_p = 2{\cdot}1$ s for $C_{10}H_8$ and 16·9 s for $C_{10}D_8$. A similar effect has been observed with the luminescence of rare earth ions when the water of hydration is replaced by D_2O. This phenomenon arises because deuteration markedly reduces $k_{isc}(T_1 \rightsquigarrow S_0)$, a point discussed on p. 92.

3.2.4 Delayed Fluorescence[17]

Delayed fluorescence, in which the luminescence decays more slowly than normal 'prompt' fluorescence from the same molecule, can arise by several mechanisms, of which the most closely investigated are triplet–triplet annihilation and thermally-activated delayed fluorescence.

Triplet–Triplet Annihilation

The fluorescent emission from a number of aromatic hydrocarbons (e.g., naphthalene, anthracene, phenanthrene) shows two components with identical emission spectra. One component decays at the rate for normal fluorescence, and the other has a lifetime approximately half that of phosphorescence. The implication of triplet species in this delayed fluorescence, suggested by the fact that the delayed emission can be induced by triplet sensitizers, has been confirmed by kinetic analysis. The accepted mechanism[18] is:

$$S_0 \xrightarrow[I]{h\nu} S_1 \quad \text{absorption} \tag{3.19}$$

$$S_1 \xrightarrow{k_f} S_0 + h\nu' \quad \text{normal fluorescence} \tag{3.20}$$

$$S_1 \xrightarrow{k_{isc}} T_1 \quad \text{intersystem crossing} \tag{3.21}$$

$$T_1 \xrightarrow{k_p} S_0 + h\nu'' \quad \text{phosphorescence} \tag{3.22}$$

$$T_1 + T_1 \xrightarrow{k_5} X \tag{3.23}$$

$$X \xrightarrow{k_6} S_1 + S_0 \quad \text{(3.23)–(3.24): triplet–triplet annihilation (spin–allowed)} \tag{3.24}$$

$$X \xrightarrow{k_7} S_0 + S_0 \quad \text{deactivation} \tag{3.25}$$

$$S_1 \xrightarrow{k_f} S_0 + h\nu' \quad \text{delayed fluorescence} \tag{3.26}$$

The crucial steps are equations (3.23) and (3.24), in which two excited triplets, on collision, redistribute their energies via the entity X so that one is promoted to S_1 and the other collapses to the ground state. It is the S_1 state so produced that is responsible for the delayed fluorescence. Although it emits (equation 3.26) with the same rate constant as prompt fluorescence (k_f), its decay is inhibited because it continues to be regenerated *via* steps (3.23) and (3.24).

Application of the steady state approximation shows that under conditions of low exciting light intensity, where $[T_1]$ will be small and terms involving $[T_1]^2$ can be neglected, the intensity of delayed fluorescence (I_{DF}) is given by:

$$I_{DF} = \frac{k_5 k_6}{k_6 + k_7}\left[\frac{k_{isc} I}{k_p(k_f + k_{isc})}\right]^2 \tag{3.27}$$

Also,

$$k_{DF} = -\frac{d(\ln I_{DF})}{dt} = -\frac{d(\ln[T_1]^2)}{dt} = -2\frac{d(\ln[T_1])}{dt}$$

so that

$$k_{DF} = 2k_p \quad \text{and} \quad \tau_{DF} = \tfrac{1}{2}\tau_p \tag{3.28}$$

These predictions, (i) that the intensity of the delayed emission should be proportional to the square of the incident light intensity and (ii) that the lifetime of the delayed fluorescence should be one-half that of phosphorescence, have been abundantly confirmed. The intermediate X in the above mechanism is clearly an excited dimer or excimer (see Chapter 4), and the reaction of equation (3.24) is the reverse of that involved in the concentration quenching of fluorescence (see Chapter 4, p. 103). For anthracene, k_6 is so large that delayed fluorescence of only the monomer is observed. With pyrene, on the other hand, k_6 is much smaller, and delayed emission derives from both the monomer and (mainly) the dimer. Delayed fluorescence from the dimer, unlike that from the monomer, is not the same as normal fluorescence.

Delayed fluorescence of this type has been detected in gases, solution, rigid glasses and even crystals. In the last two phases, where the diffusion of triplets is difficult or impossible, the emission probably depends on exciton migration—a form of energy transfer in which the energy of excitation 'hops' from molecule to molecule. Ultimately two adjacent triplets are formed and delayed emission ensues.

Thermally-activated Delayed Fluorescence

This emission is characterized by the following features:

(i) The prompt and delayed fluorescence emission spectra are identical.

(ii) $\tau_{DF} = \tau_p$ (contrast triplet–triplet annihilation).

(iii) $I_{DF} \propto I$ (intensity proportional to absorbed light intensity, contrast triplet–triplet annihilation).

(iv) The ratio of the intensities of delayed fluorescence and phosphorescence decreases exponentially with the singlet–triplet splitting and with the reciprocal of the absolute temperature, i.e. $I_{DF}/I_p \propto e^{-(E_S - E_T)/RT}$. No emission is observed at low temperatures.

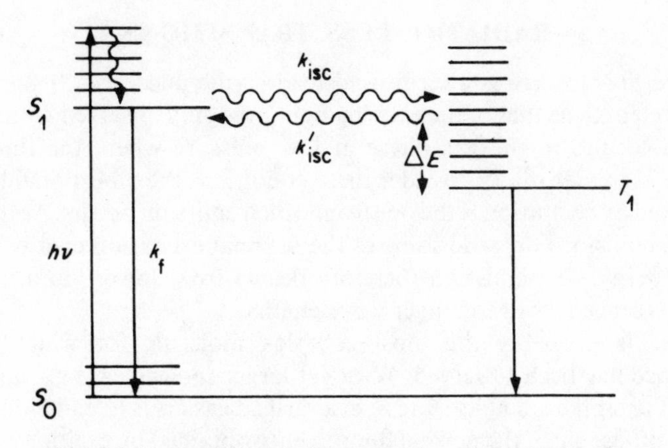

Figure 3.18. Energy levels for thermally-activated delayed fluorescence

These facts are explained by the mechanism depicted in Figure 3.18. Light absorption followed by intersystem crossing and vibrational relaxation gives triplets in their zero-point vibrational level. Thermal activation through the energy gap ΔE followed by reverse intersystem crossing ($T_1 \rightsquigarrow S_1$) gives excited singlets, which then fluoresce. It follows that:

$$I_p = k_p[T_1]$$

and

$$I_{DF} = \theta_f \,.\, k'_{isc}[T_1]\,e^{-\Delta E/RT}$$

where θ_f is the quantum efficiency of fluorescence, whence

$$\frac{I_{DF}}{I_p} = \frac{\theta_f \,.\, k'_{isc}}{k_p} \,.\, e^{-\Delta E/RT} \tag{3.29}$$

On this mechanism, it would be expected that the experimentally measured activation energy ΔE should correspond to the singlet–triplet splitting obtained from spectroscopic data, and within experimental error this is found to be so. It has also been found that $k_{isc} \simeq k'_{isc}$.

It should be noted that thermally-activated delayed fluorescence seems to be largely confined to certain dyestuffs (eosin, fluorescein, acriflavine and proflavine) in which the singlet–triplet splitting is small (20–40 kJ mol^{-1}, 5–10 kcal mol^{-1}). With aromatic hydrocarbons the magnitude of the (π, π^*) splitting prohibits this mechanism for delayed fluorescence.

The importance of delayed fluorescence is that, in exploiting the extremely high sensitivity associated with emission spectroscopy, it provides a powerful tool for examining the behaviour of relatively low concentrations of triplets in solutions. The major alternative technique, flash photolysis, requires a high concentration of triplets, and this can introduce difficulties. Also, delayed emission gives information about the rates of all three intersystem crossings and permits triplet–triplet quenching to be observed directly.

3.3 RADIATIONLESS TRANSITIONS[19]

Resonance fluorescence is a term used to describe fluorescent emission of the same wavelength as that of the exciting light. It is only observed with atoms and simple molecules in the gas phase at low pressure where the time between collisions is greater than τ_f. Under these conditions the vibrational level which is populated by excitation is the one from which emission occurs. As the pressure increases, collisions degrade some of the original excitation energy into translational energy. The emission therefore occurs from lower vibrational levels, and fluorescence moves to longer wavelengths.

Benzene is probably the most complex molecule for which resonance fluorescence has been observed. With yet larger molecules (e.g., naphthalene), excitation to a higher singlet state S_n in a rarified gas leads to emission similar to, but more diffuse than, the normal fluorescence obtained by exciting S_1. It seems, then, that sufficiently large molecules in their S_n states undergo a transition to S_1 before luminescing. Since this transition ($S_n \rightsquigarrow S_1$) is not accompanied by photon emission, it is referred to as a *radiationless* or *non-radiative transition.* That is occurs in low-pressure gases indicates that it is non-collisional and therefore probably intramolecular. Since energy must be conserved, it follows that in this case the radiationless transition must be from S_n to an isoenergetic level of S_1. Phenomenologically the process can be represented by a horizontal (wavy) line on a Jablonski diagram (Figure 3.19).

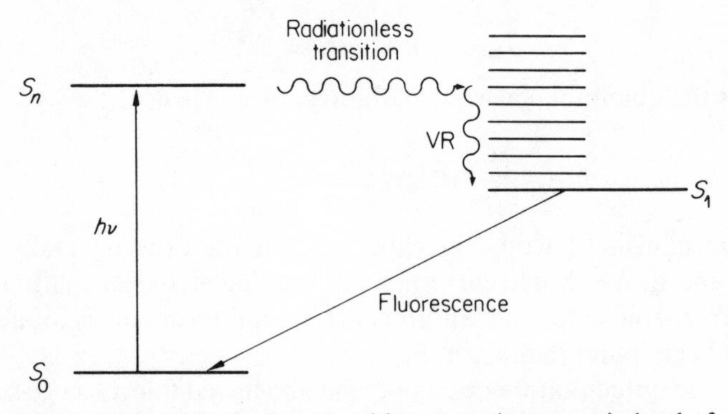

Figure 3.19. Radiationless transition to an isoenergetic level of a second state—horizontal wavy line on a Jablonski diagram

As indicated above, the emissive behaviour is a function of the size of molecules. It also depends on the environment, i.e. on whether the excited species is in solution or trapped in a dense medium or is in a rarified gas. This variation in emission reflects changes in the rates of radiationless transitions. A unified quantum-mechanical theory[19,20] is being rapidly developed at the moment which seeks to account for all the diverse phenomena associated with radiationless processes. It is impossible in a book of this sort to deal adequately with this theory. Rather we shall concern ourselves with radiationless phenomena exhibited by large molecules (i.e., benzene or larger) in condensed phases, because such systems are of greatest interest to organic photochemists and because the theoretical treatment is simpler.

3.3.1 Summary of Experimental Data on Large Molecules in Dense Media

The basic data emerging from many measurements is:

(i) Emission almost invariably occurs from S_1 or T_1, independent of the state which is initially excited (Kasha's Rule).†

(ii) ϕ_f does not depend on which state (S_1, S_2, S_3 etc.) is first excited (Vavilov's Law). This implies that very rapid internal conversion occurs between upper excited singlet states.

(iii) $\phi_f < 1$ and often $\phi_f \ll 1$.

(iv) The observed values of the fluorescent lifetime τ_f are less than the radiative lifetime τ_0 calculated from the oscillator strength (see Chapter 2, p. 14).

(v) The rates of radiationless transitions conform to an '*energy gap law*', which will be discussed in more detail later, and according to which the rate of radiationless transitions falls very rapidly with increase in the energy difference between the $v = 0$ levels of the states concerned.

3.3.2 A Simple Model

Some understanding of radiationless processes in polyatomics in condensed phases may be obtained from the (oversimplified) model represented by Figure 3.20.

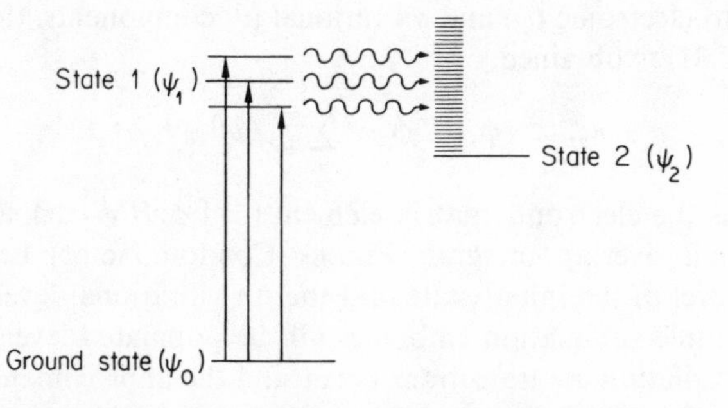

Figure 3.20. A model for radiationless transitions in polyatomics in condensed phases

The molecule has a ground state (ψ_0), an excited state (ψ_1) into which it can be excited, and another excited state (ψ_2) of lower energy (either triplet or lower singlet, not excluding the ground state) and with a dense manifold of states provided by vibrational and rotational sublevels, some of which are isoenergetic with those of ψ_1. The essential near-degeneracy between the levels of ψ_1 and ψ_2

† The notable exceptions are azulene and its derivatives, which fluoresce strongly from S_2 and whose emission from S_1 is so weak that it has only recently been detected, and pyrene, 3,4-benzpyrene and some thiocarbonyl compounds.

can be achieved, if necessary, by adding in small amounts of lattice energy from the medium.† In other words, solvent perturbations broaden the sublevels.

Excitation leads to population of a vibrational level or levels $v' = m$, from which a non-radiative transition to the quasi-continuum of state 2 occurs. These vibrational levels may be those initially populated, or, if k_{VR} the rate constant for vibrational relaxation is much greater than k_{nr} the rate constant for non-radiative transition, then the $v' = 0$ level of ψ_1 is that from which the radiationless transition occurs. In any case, the transition can be treated by time-dependent perturbation theory (see Chapter 2), a perturbation H' inducing a time-dependent evolution of the system from an initial state (ψ_1) into a final one (ψ_2). For internal conversion H' arises from electrostatic interactions between the electrons and nuclei, and for intersystem crossing H' is the spin–orbit interaction. Therefore internal conversion and intersystem crossing are treated similarly. Subject to certain conditions,[21] the rate constant k_{nr} for the non-radiative transition from each populated level of state 1 is then given by the Fermi golden rule:

$$k_{nr} = \frac{2\pi}{h}\langle\psi_1|H'|\psi_2\rangle^2\rho \tag{3.30}$$

where ρ is the state-density factor which describes the number of states in the quasi-continuum isoenergetic with the levels of state 1 from which the radiationless transition occurs, and $\langle\psi_1|H'|\psi_2\rangle$ is a matrix element giving the energy of the interaction between the initial and final states induced by the perturbation H'. Invoking the Born–Oppenheimer approximation and factorizing the wavefunctions into electronic (ϕ) and vibrational (θ) components, the approximate expression (3.31) is obtained.

$$k_{nr} \propto \langle\phi_1|H'|\phi_2\rangle^2 \sum_i \sum_j \langle\rho\theta_{1i}|\theta_{2j}\rangle^2 \tag{3.31}$$

$\langle\phi_1|H'|\phi_2\rangle$ is the electronic matrix element ($\equiv \int \phi_1 H' \phi_2 \, d\tau$), and $\langle\theta_{1i}|\theta_{2j}\rangle$ is the vibrational overlap integral (Franck–Condon factor) between the ith vibrational level of the initial state and the jth vibrational level of the second state. The double summation embraces all the populated levels of the state 1 from which radiationless transitions occur and the approximately isoenergetic levels of state 2 to which the transition occurs, and ρ is the state-density factor weighting each Franck–Condon term by the number of sublevels of each vibrational level.

Non-radiative phenomena in large molecules will now be discussed in the light of this equation.

† In small molecules with widely spaced vibrational levels in the gas phase at pressures where collisions are unimportant, the required degeneracy of the levels will occur only rarely, and radiationless transitions are very unlikely, so that $\phi_f \sim 1$. For sufficiently large molecules, however, with a large number of vibrational levels, ψ_2 becomes a quasi-continuum and degeneracy is readily achieved. Thus radiationless transitions would be expected to occur in large isolated molecules, as is observed to be the case.

3.3.3 The Electronic Matrix Element

This is the non-radiative counterpart of the electronic transition moment (see Chapter 2) and, like the latter, will have very small values unless the initial and final orbitals overlap effectively. Symmetry may, on occasion, make this integral zero, leading to a forbidden radiationless transition. It is possible to analyse such situations with the aid of group theory (see Appendix). For intersystem crossing, H' is the spin–orbit coupling operator H_{SO}, which can be resolved into three perpendicular components which transform like rotations R_x, R_y and R_z in the group character tables.

Consider the intersystem crossing $^1(n, \pi^*) \rightsquigarrow {}^3(n, \pi^*)$. Since both states are (n, π^*) they will have the same spatial wavefunction (say φ). The electronic matrix element is then $\langle\varphi|H_{SO}|\varphi\rangle$. The direct product $\Gamma\varphi \times \Gamma\varphi$ of the irreducible representations of φ will belong to the totally symmetric irreducible representation. Rotations in the point groups corresponding to the vast majority of molecules of interest to organic photochemists never belong to the totally symmetric irreducible representation. The triple product $\Gamma\varphi \times \Gamma\varphi \times \Gamma H_{SO}$ will therefore not be totally symmetric, the matrix element will be zero, $k_{nr} = 0$, and the transition will be forbidden. The same arguments apply to the intersystem crossing $^1(\pi, \pi^*) \rightsquigarrow {}^3(\pi, \pi^*)$. However, for $^1(n, \pi^*) \rightsquigarrow {}^3(\pi, \pi^*)$, the direct product $\Gamma\varphi_1 \times \Gamma\varphi_2$ is not totally symmetric, so that radiationless transitions may be allowed.

This is the theoretical basis of *El Sayed's selection rules* for intersystem crossing:

Allowed: $^1(n, \pi^*) \leftrightsquigarrow {}^3(\pi, \pi^*)$; $^3(n, \pi^*) \leftrightsquigarrow {}^1(\pi, \pi^*)$

Forbidden: $^1(n, \pi^*) \leftrightsquigarrow {}^3(n, \pi^*)$; $^1(\pi, \pi^*) \leftrightsquigarrow {}^3(\pi, \pi^*)$

Processes forbidden under these rules still occur, but with rate constants 10^{-2}–10^{-3} times those for allowed intersystem crossings.

The preceding discussion, based on symmetry, has implicitly assumed the validity of the Born–Oppenheimer approximation. To be more accurate, vibronic wavefunctions ψ should be used and the value of integrals of the form $\langle\psi_i|H_{SO}|\psi_f\rangle$ computed. Since H_{SO} is an electronic operator and since the Born–Oppenheimer approximation is a close one, the values of the matrix elements $\langle\psi_i|H_{SO}|\psi_f\rangle$ and $\langle\varphi_i|H_{SO}|\varphi_f\rangle$ will be similar. However, the symmetries of ψ and φ will, in general, be different, because the symmetry of the former contains contributions from the symmetry of the vibrational components, whereas the latter (Born–Oppenheimer) wavefunction does not. Thus, when the orbital matrix element is zero, the vibronic matrix element will often be non-zero though small (cf. discussion of radiative vibronic coupling, Appendix and Chapter 2, section 2.1.8). In other words, rigorous symmetry restraints can be relaxed by including vibrational terms (vibronic coupling). There is evidence that much of the observed non-radiative decay ($S_1 \rightsquigarrow T_1$) in aromatics occurs *via* vibronic spin–orbit coupling.

The well known heavy atom effects in intersystem crossing (see Chapter 2, p. 24 and Chapter 3, p. 82) are, of course, another manifestation of the role of the electronic matrix element, which becomes magnified because $H' = H_{SO}$ is greater in systems containing heavy atomic nuclei. This leads to increased values of k_{isc} with the consequences discussed earlier.

For internal conversion, the electronic matrix element is $\langle \varphi_1 | H_{ic} | \varphi_2 \rangle$ and H_{ic} is the nuclear kinetic energy operator.† Since H_{ic} belongs to the totally symmetric irreducible representation, it is immediately apparent that the integrand will be totally symmetric and the matrix element non-zero only if φ_1 and φ_2 belong to the *same* irreducible representation. What this means is that, rigorously, only states of the same symmetry should internally convert (contrast intersystem crossing), and hence, since S_1 and S_0 must have different symmetries, the internal conversion $S_1 \rightsquigarrow S_0$ is forbidden. That this phenomenon occurs is again due to vibronic coupling, for the crossing from S_1 must be to very high vibrational levels of S_0 and some of these will have the same vibronic symmetries as those of S_1.

3.3.4 Vibrational Overlap Integral (Franck–Condon Factor)

Before proceeding, the reader will find it helpful to read the material on vibrational overlap integrals in Chapter 2 (section 2.1.5).

Radiationless transitions can be visualized by thinking of the intersection of potential energy surfaces. A molecule on the potential energy surface corresponding to state 1 'crosses' at the point of intersection (X) to the potential energy surface associated with state 2. Consider the disposition of energy surfaces‡ in Figure 3.21. At the internuclear separation corresponding to the intersection, state 1 and state 2 have the same energy and the same internuclear distance. A molecule in the level AX of state 1, when it arrives at X, has merely to change the quantum numbers of one of its electrons (to rearrange the motion of one of its electrons) to be in state 2 in the level BX. Since X is a turning point

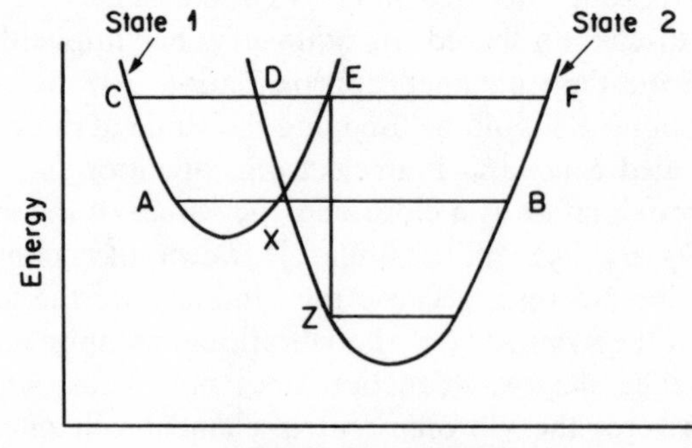

Figure 3.21. The Franck–Condon principle and radiationless transitions

† Although H_{ic} operates only on nuclei, it nevertheless affects electronic wavefunctions because of their dependence upon nuclear coordinates.

‡ The complete representation of the energy of a molecule of N atoms requires a hypersurface in ($3N$-6) dimensions. For the purpose of visualization it is common practice to take a section through the hypersurface corresponding to changes in only one internuclear distance—in other words to treat polyatomics as though they are diatomics (this can be misleading).

in the vibrations A–X–A and B–X–B, where the kinetic energy is zero and the system is momentarily at rest, more time is available at this point than at any other for the system to cross from one state to the other. The process becomes irreversible if the molecule having arrived in the level BX undergoes rapid vibrational relaxation.

Now consider the possibility of a radiationless transition from level CE. If the transition occurs when the molecule is at the turning point E, then virtually instantaneously there must either be a change in the nuclear coordinates to arrive at points D or F in state 2, or a change in kinetic energy (equal to EZ) if the internuclear separation remains constant. Similar arguments apply to transitions occurring at (say) point D. Such rapid changes are forbidden by the Franck–Condon principle, which applies equally to radiative and non-radiative processes.

It is desirable for later purposes to recast the above classical argument into quantum-mechanical terms based on vibrational overlap integrals. In Chapter 2 vertical (radiative) transitions were so analysed. We now wish to understand the impact of such considerations on horizontal (non-radiative) transitions. If θ_1 and θ_2 are vibrational wavefunctions associated with two electronic states designated 1 and 2, then the probability of crossing between the states is proportional to $\int \theta_1\theta_2 \, d\tau_N$. Compare the situation of Figure 3.22, where the zero-point levels of the two states have approximately the same energy, with that of Figure 3.23, where there is a considerable gap between the zero-point levels.

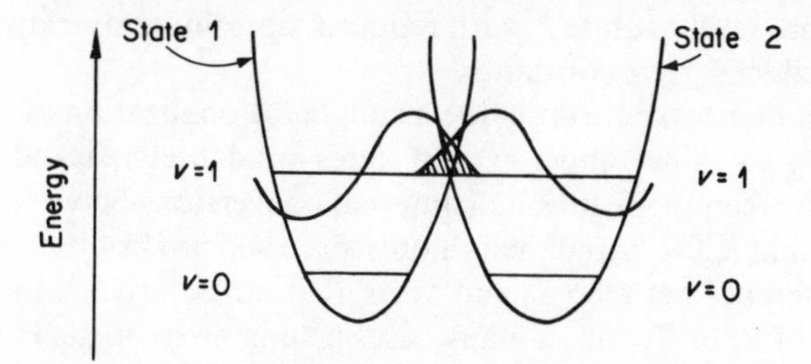

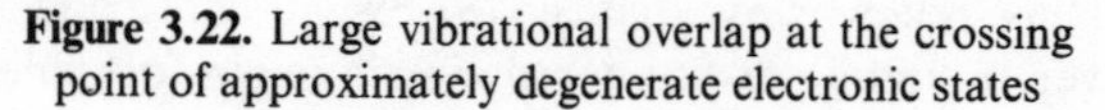

Figure 3.22. Large vibrational overlap at the crossing point of approximately degenerate electronic states

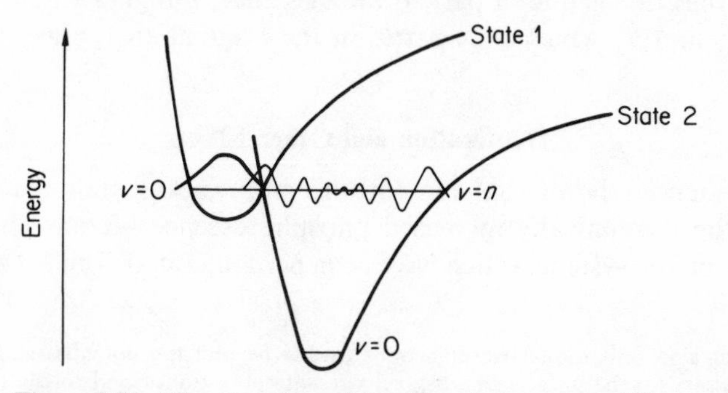

Figure 3.23. Vibrational overlap with a large energy difference between states

It is immediately apparent that the vibrational overlap integral at the crossing point in Figure 3.22 is of significant dimensions, but because of the rapidly oscillating character of the $v = n$ function in Figure 3.23, the positive contributions to the vibrational overlap integral are largely cancelled by the negative contributions, so that the integral is very small.†

A molecule in each of the populated levels of state 1 has therefore a definite probability of crossing to each of the many approximately isoenergetic levels of state 2, and the total probability of crossing is given by the double summation term of equation (3.31). The effect of this term is seen in the energy gap law, in the effect of deuteration on the rates of radiationless transitions, and in certain other phenomena now to be discussed.

3.3.5 Energy Gap Law

There is an inverse correlation between the rates of non-radiative transitions involving the lowest states of similar molecules and the difference in energy between the $v = 0$ levels of the states involved. In other words, the smaller the energy gap the bigger the rate. This is shown in Figures 3.24 and 3.25, where the logarithms of the rate constants for intersystem crossing ($T_1 \rightsquigarrow S_0$) and for internal conversion ($S_1 \rightsquigarrow S_0$) for a series of aromatic hydrocarbons are plotted against the energy gap.

The energy gap law is readily understood by noting that as the gap increases the radiationless transition from a given level of state 1 will be to an increasingly high vibrational level of state 2, with reduced vibrational overlap and a correspondingly reduced rate constant.

The law can be invoked to provide a simple rationalization of Kasha's Rule and Vavilov's Law. Since upper excited states are densely packed, i.e. since the energy gaps between them are small, internal conversions between them will be very rapid, so that $k_{ic} \gg k_f$ and their fluorescence will not be observed. However, the energy difference between S_0 and S_1 or T_1 is much larger, and radiationless depopulation of S_1 or T_1 will in many cases be unable to quench emission from these states. It is noteworthy that $k_{isc}(S_1 \rightsquigarrow T_1)$ is frequently very large ($\sim 10^{10}\ s^{-1}$), even though intersystem crossing is a spin-forbidden process, and it may be that this is due in part to the existence of higher triplet states lying between S_1 and T_1 which thus partition the original energy gap into smaller ones.

3.3.6 Deuteration and Other Effects

Further evidence relating to the importance of Franck–Condon factors may be found in the dramatically increased phosphorescence lifetimes of the triplet states of aromatic systems which have been perdeuterated. Table 3.4 exemplifies this point.

† Note that since vibrational wavefunctions extend beyond the potential surfaces, it is not actually necessary for the surfaces to cross, i.e. a radiationless transition between the $v = 0$ levels in Figure 3.22 is still possible, though with strongly reduced probability (quantum mechanical 'tunnelling').

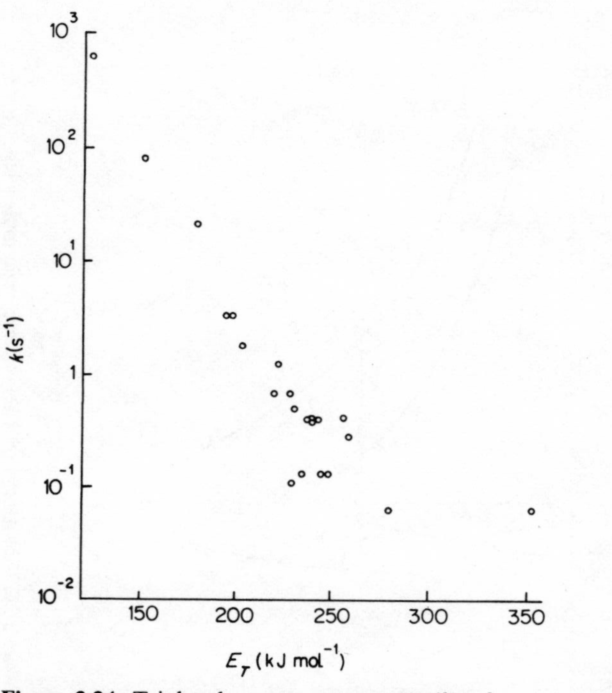

Figure 3.24. Triplet decay rate constants (k) of aromatic hydrocarbons plotted against the triplet energy (E_T). (From W. Siebrand, in A. B. Zahlan (ed.), *The Triplet State*, (1967), reproduced by permission of Cambridge University Press)

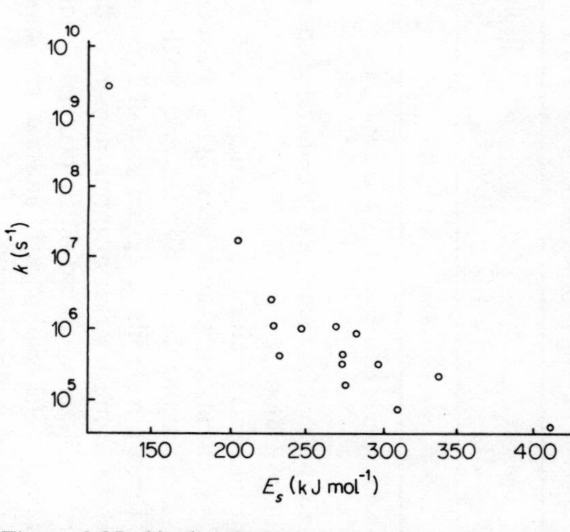

Figure 3.25. Singlet decay rate constants (k) of aromatic hydrocarbons plotted against the singlet energy (E_S). (From data in J. B. Birks, *Photophysics of Aromatic Molecules*, (1970), Wiley)

Table 3.4. Effect of deuteration on phosphorescent lifetimes of aromatic hydrocarbons[a]

Compound	Benzene C_6H_6	C_6H_5D	C_6D_6	Naphthalene $C_{10}H_8$	$C_{10}D_8$
τ_p(s)	5·75	8·90	12·0	2·51	21·7

[a] Measured in 3-methylpentane glass, except $C_{10}D_8$ in E.P.A. glass.

It has also been shown that for incompletely deuterated compounds, τ_p depends upon the position of deuteration.

The deuterium effect is explained on the assumption that k_{nr} is largely determined by the C—H vibrations because, being of high frequency, they are widely spaced. The C—D vibrations are of lower frequency, so that the levels are more closely packed. Hence a molecule in a given level of T_1 crosses to a vibrational level of S_0 which has a much larger quantum number in the deuterated compound. $k_{isc}(T_1 \rightsquigarrow S_0)$ is thus reduced in the deuterated compound, and τ_p is increased accordingly. The result clearly indicates the important influence of C—H stretching modes in determining k_{nr}.

There is another effect which depends on Franck–Condon factors. With large and rigid molecules (e.g., aromatics) there is little change of geometry on excitation, so that the minimum in the S_1 surface is likely to be only slightly displaced with respect to that of S_0. In such cases, as inspection of Figure 3.26 reveals, internal conversion from the zero-point level of S_1 to S_0 will be very slow because the vibrational overlap integral will be very small. This means that fluorescence can compete with radiationless depopulation of S_1, and hence the rule that rigid systems tend to fluoresce.

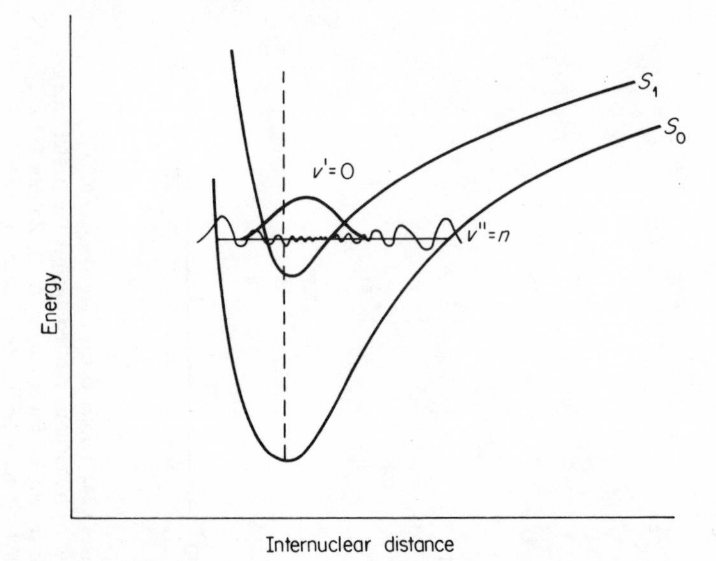

Figure 3.26. Vibrational overlap in large, rigid molecules

In condensed phases the rate of vibrational relaxation is likely to be much larger than k_{nr}, so that radiationless transitions would be expected to occur exclusively from the $v = 0$ level of state 1 (except in so far as thermal activation will induce a Boltzmannian population of higher vibrational levels). That this is indeed the case is shown by innumerable observations that the nature and lifetime of the emission from large molecules in such phases is independent of the wavelength of the exciting light and therefore of which vibrational levels are initially populated. Since molecules in the various vibrational levels of state 1 have different probabilities of crossing to state 2, it follows that the emission parameters will be wavelength-dependent in a low pressure gas phase. This has been found to be so. Recent work has focussed upon exciting individual vibrational levels and determining relative values of k_{nr}.[22] A theoretical model[23] has been developed for treating the non-radiative decay of single vibrational levels of large molecules, using state-density-weighted Franck–Condon factors, which gives excellent agreement with experimental observations for benzene and its derivatives in the gas phase.

3.3.7 Fluorescence vs. Phosphorescence

Whether a molecule fluoresces or phosphoresces or does neither, depends on the values of k_f, k_p, ${}^1k_{isc}(S_1 \rightsquigarrow T_1)$ and ${}^1k_r$ and ${}^3k_r$ (the rate constants for photochemical change in excited singlet and triplet states), Figure 3.27. Neglecting

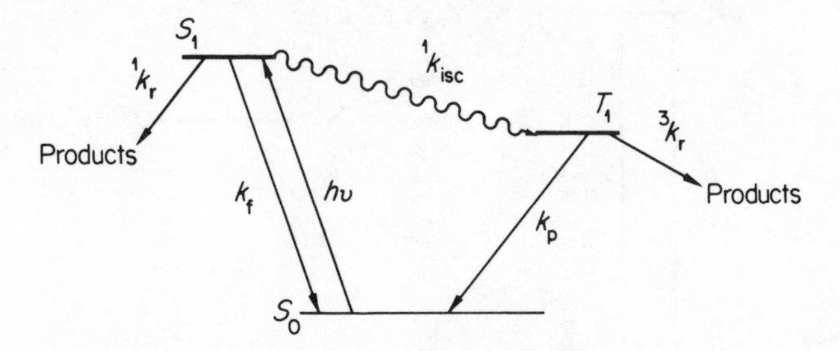

Figure 3.27. A kinetic scheme for deriving quantum yields of fluorescence and phosphorescence

internal conversion ($S_1 \rightsquigarrow S_0$) and intersystem crossing ($T_1 \rightsquigarrow S_0$) because their rate constants are too small to affect the argument, it is immediately apparent that an excited species can revert to the ground state only by photochemical reaction or luminescence. In other words, absence of luminescence means that rapid photochemical changes are occurring. However, if the photoproducts are highly labile entities which revert rapidly to starting material, then no overall change will be observed. This point is taken up in the next section.

Applying equations (3.3) and (3.10) to the situation of Figure 3.27, the quantum yields of fluorescence and phosphorescence can be expressed as:

$$\phi_f = \frac{k_f}{{}^1k_r + k_f + {}^1k_{isc}} \tag{3.32}$$

and

$$\phi_p = \frac{k_p}{^3k_r + k_p} \cdot \frac{^1k_{isc}}{^1k_r + k_f + {^1k_{isc}}} \tag{3.33}$$

From these expressions it is apparent that fluorescence will be observed only if $k_f \nless (^1k_r + {^1k_{isc}})$, and phosphorescence will be observed only if both $k_p \nless {^3k_r}$ *and* $^1k_{isc} \nless (^1k_r + k_f)$. Furthermore, $^1k_{isc}$ imposes limits on the processes open to S_1. Clearly, if $^1k_{isc} \gg {^1k_r}$, then any photochemical changes involving S_1 can only proceed with a very small quantum yield. Equally if $^1k_{isc} \gg k_f$, no fluorescence will be detectable.

Since $^1k_{isc}$ is normally at least comparable with k_f, triplets are always produced, and phosphorescence must occur unless $^3k_r \gg k_p$. Since k_p is small this condition is often satisfied, and phosphorescence is by no means a widespread phenomenon among organic molecules as a whole.

These considerations, illustrated in Figures 3.28 and 3.29, show why benzophenone fails to fluoresce strongly while aliphatic ketones both fluoresce and phosphoresce. The essential point is that $^1k_{isc}$ is much greater for benzophenone because of the operation of El Sayed's selection rules.

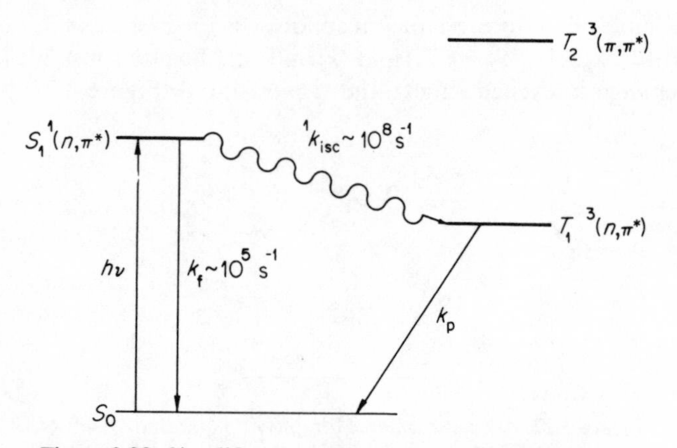

Figure 3.28. Simplified state diagram for aliphatic ketones

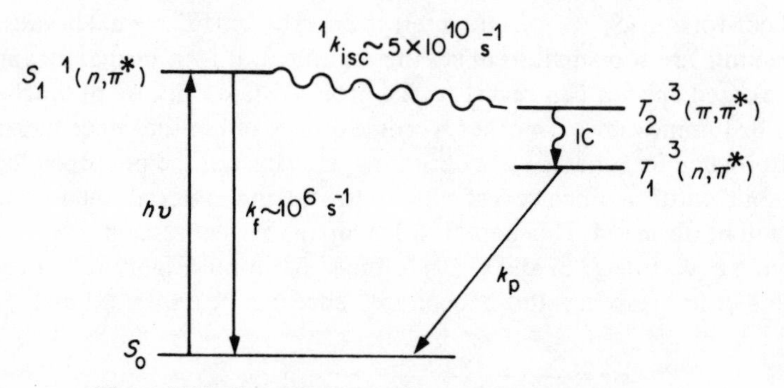

Figure 3.29. Simplified state diagram for benzophenone

Emission Rate Constants

Since $k_f = A_{ul}$, the Einstein coefficient of spontaneous emission, and A_{ul} is related to B_{ul}, the discussion in Chapter 2 on the intensity of absorption applies equally to k_f. Hence k_f will be large (of the order of 10^7–10^8 s^{-1}) for allowed or partially allowed $\pi \rightarrow \pi^*$ transitions and will be very much smaller ($\sim 10^5$ s^{-1}) for forbidden $n \rightarrow \pi^*$ transitions. Therefore, other things being equal, aromatic hydrocarbons would be expected to fluoresce more strongly than simple ketones.

With respect to phosphorescence, in Chapter 2 (p. 21) the intensities of singlet–triplet transitions were related to the extent that H_{SO}, the spin–orbit coupling operator, mixes into nominal triplet and singlet states small amounts of the state of the other multiplicity (see equation 2.12). If we can neglect the mixing of triplet states into S_0 because of the large energy gap ($E_T - E_{SO}$), then the rate of phosphorescence from a given nominal triplet level will be correlated to the amount of the singlet component of that triplet state. These concepts are now applied to aliphatic ketones and aromatic hydrocarbons.

In aliphatic ketones, T_1 is $^3(n, \pi^*)$ and in the C_{2v} point group; $\Gamma(n, \pi^*) = A_2$ (see Appendix). Bearing in mind that the components of H_{SO} transform like rotations, examination of the C_{2v} character table shows that H_{SO} mixes the triplet (n, π^*) state of the carbonyl group with the singlet states $^1(\sigma, \pi^*)(^1B_1)$, $^1(n, \sigma^*)(^1B_2)$ and $^1(\pi, \pi^*)(^1A_1)$. It follows that the nominal $^3(n, \pi^*)$ state of aliphatic ketones will have significant singlet character, leading to a relatively large rate for transition to the singlet ground state and hence to a short triplet lifetime.

Conversely, in benzene where all the lowest states are (π, π^*), appeal to the character table for the D_{6h} point group shows that the lowest triplet T_1, which belongs to the B_{1u} representation, cannot be mixed by H_{SO} with any of the singlet (π, π^*) states. Thus T_1 would be expected to be a virtually pure triplet state except in so far as vibronic spin–orbit coupling (Chapter 3, p. 89) relaxes these symmetry restraints. Radiative collapse to the ground state is thus expected to be very slow. To the extent that similar arguments apply to other aromatic hydrocarbons, it can be understood why for aromatic hydrocarbons τ_p is of the order of 10 s and for aliphatic ketones $\tau_p \sim 10^{-3}$ s.

3.3.8 Radiationless Transitions, Isomerization and Photochemistry

A photostable molecule, excited into S_1, must either fluoresce or undergo intersystem crossing or internal conversion. This implies that:

$$\phi_f + \phi_{ic} + \phi_T = 1 \quad (\text{note that } \phi_T \equiv \theta_{isc} \geqslant \phi_p) \tag{3.34}$$

Because of the large energy gap between S_0 and S_1, k_{ic} is expected to be small relative to k_f and k_{isc}, and ϕ_{ic} is expected to be negligible. Hence $(\phi_f + \phi_T) \sim 1$, and this is found to be the case for many molecules (see Table 3.5). In other systems however, $(\phi_f + \phi_T) \ll 1$. This must mean that k_{ic} and/or $^3k_{isc}(T_1 \rightsquigarrow S_0)$ is large in spite of the energy gap law, and it is tempting to ascribe the enhanced

Table 3.5. Quantum yields for fluorescence and triplet formation

Compound	ϕ_f	$\phi_T(\equiv \theta_{isc})$
Naphthalene	0·21	0·71
Anthracene	0·33	0·58
Phenanthrene	0·14	0·70
Triphenylene	0·09	0·89
1-Methoxynaphthalene	0·53	0·46
9-Phenylanthracene	0·45	0·505
Fluorescein	0·92	0·05
Eosin	0·19	0·71
Erythrosin	0·02	1·07
Pyridazine	0·01	0·00
Pyrimidine	0·0058	0·14
Pyrazine	0·0006	0·30

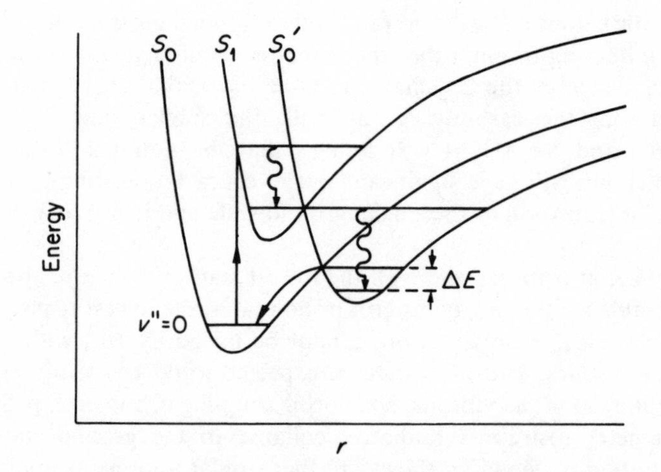

Figure 3.30. A possible role of intermediates in internal conversion

rate of radiationless decay found in such systems to the formation of unstable isomers.[24] Figure 3.30 depicts the situation.

Internal conversion from S_1 to the ground state isomer S_0' will be much faster than to S_0 (Energy gap law). Vibrational relaxation sends the molecule into the zero-point level of S_0', and if the energy barrier ΔE between S_0' and S_0 is small enough to be surmounted rapidly by thermal activation, then all that will be observed is an enhanced rate of internal conversion. If ΔE is larger than a few kT (where k is Boltzmann's constant), then the molecule will remain in S_0' as a metastable isomer. One system exhibiting this phenomenon is biacetyl, which on $n \longrightarrow \pi^*$ excitation in water or methanol gives rise to the enol (3.35), which of course reverts to biacetyl. Another is benzene, which on irradiation is transformed into benzvalene (3.36), which slowly reverts to benzene at room temperature.

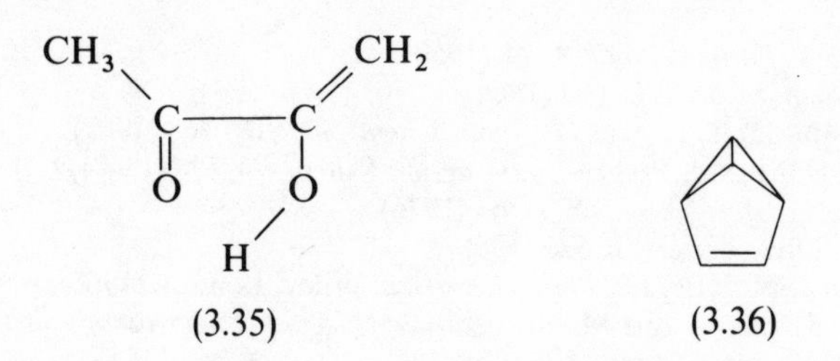

(3.35) (3.36)

At this point we should pause and take stock. Radiationless transitions occurring between states of the *same* molecule have been discussed. Formally:

$$\left.\begin{array}{r} {}^1M^{**} \rightsquigarrow {}^1M^* \\ {}^1M^* \rightsquigarrow {}^1M \\ {}^3M^{**} \rightsquigarrow {}^3M^* \end{array}\right\} \text{internal conversion} \qquad \left.\begin{array}{r} {}^1M^* \rightsquigarrow {}^3M^* \\ {}^3M^* \rightsquigarrow {}^1M \end{array}\right\} \text{intersystem crossing}$$

Then the concept of radiationless transitions between states of *isomeric* molecules was introduced:

$${}^1M^* \rightsquigarrow {}^1M' \rightsquigarrow {}^1M$$

There seems to be no difference in principle between any of the above photophysical processes and the photochemical processes:

$${}^1M^* \rightsquigarrow {}^1X \quad \text{or} \quad {}^3M^* \rightsquigarrow {}^1X$$

which involve radiationless transitions between states of *different* molecules. Radiationless transitions and photochemical reactions both involve a non-radiative electron rearrangement whereby an excited state decays into a state of lower energy. This implies a conversion of electronic energy into vibrational energy.[25] The difference between the two classes of phenomena seems to be this, that in radiationless transitions the electronic decay is into quantized vibrational levels, whereas in photochemical reactions the decay is into continuum states, for bonds are broken and/or rearranged. The electronic-vibrational energy transfer is often highly selective in that only certain vibrational modes are excited, as witnessed by the orbital symmetry rules (see Chapter 6) and the frequent observation that bonds to hydrogen, usually the strongest in the molecule, are those broken photochemically. With the current rapidly developing theoretical situation it may ultimately become possible to integrate photochemical reactions into the theory of radiationless transitions, but, for the time being, other concepts are used with considerable success in rationalizing the course of such reactions.

REFERENCES

1. E. J. Bowen (ed.), *Luminescence in Chemistry*, Van Nostrand, London (1968).
2. R. S. Becker, *Theory and Interpretation of Fluorescence and Phosphoresence*, Wiley, London, (1969).
3. M. Kasha, *Discuss. Faraday Soc.*, **9**, 14 (1950).
4. C. A. Parker, *Chem. in Brit.*, **2**, 160 (1966).
5. A. P. Marchetti and D. R. Kearns, *J. Amer. Chem. Soc.*, **89**, 768 (1967).
6. J. B. Birks and I. H. Munro, *Progress in Reaction Kinetics*, **4**, 277 (1967); L. J. Clinelove and L. A. Shaver, *Analyt. Chem.*, **48**, A364 (1976).
7. E. Drent, *Chem. Phys. Letters*, **2**, 526 (1968).
8. J. G. Calvert and J. N. Pitts, Jr., *Photochemistry*, Wiley, London (1966), p. 800.
9. S. P. McGlynn, T. Azumi and M. Kinoshita, *Molecular Spectroscopy of the Triplet State*, Prentice-Hall, New Jersey (1962), p. 204.
10. W. T. Stacey and C. E. Swenberg, *J. Chem. Phys.*, **52**, 1962 (1970); R. G. Bennett and P. J. McCartin, *ibid.*, **44**, 1969 (1966); B. Stevens and M. F. Thomasz, *Chem. Phys. Letters*, **1**, 549 (1968).
11. C. A. Parker and T. A. Joyce, *Chem. Commun.*, 749 (1968).
12. W. D. Bellamy and A. G. Tweet, *Nature*, **197**, 482 (1963); W. H. Melhuish and R. Hardwick, *Trans. Faraday Soc.*, **58**, 1908 (1962).
13. Ref. 9, Chapter 2.
14. J. K. S. Wan, *Adv. Photochem.*, **9**, 1 (1974); M. Gueron in A. A. Lamola (ed.), *Creation and Detection of the Excited State*, Volume 1, Dekker, New York (1971), p. 303.
15. S. R. La Paglia, *J. Mol. Spectr.*, **7**, 427 (1961).
16. Data extracted from ref. 2, p. 139.
17. C. A. Parker, *Adv. Photochem.*, **2**, 305 (1964).
18. C. A. Parker and C. G. Hatchard, *Proc. Roy. Soc.*, **A269**, 574 (1962).
19. J. Jortner, S. A. Rice and R. M. Hochstrasser, *Adv. Photochem.*, **7**, 149 (1969); M. A. El-Sayed, *Chem. Rev.*, **77**, 793 (1977).
20. K. Freed, *Fortschr. chem. Forsch.*, **31**, 105 (1972); D. Phillips in *Chem. Soc. Spec. Reports, Photochem.*, volumes 1–8 (1970–1977); B. R. Henry and W. Siebrand in J. B. Birks (ed.), *Organic Molecular Photophysics*, volume 1, Wiley, London (1973), p. 153.
21. See D. Phillips, *Chem. Soc. Spec. Reports, Photochem.*, **4**, 59 (1973).
22. An excellent review is by D. Phillips and K. Salisbury, *Chem. Soc. Spec. Reports, Photochem.*, **4**, 228 (1973); see also E. K. C. Lee, *Accounts Chem. Res.*, **10**, 319 (1977).
23. D. F. Heller, K. F. Freed and W. M. Gelbart, *J. Chem. Phys.*, **56**, 2309 (1972).
24. D. Phillips, J. Lemaire, C. S. Burton and W. A. Noyes, Jr., *Adv. Photochem.*, **5**, 329 (1968).
25. G. S. Hammond, *Adv. in Photochem.*, **7**, 373 (1969).

Chapter 4

Quenching of Excited States

4.1 INTRODUCTION

A substance which accelerates the decay of an electronically excited state to the ground state or to a lower electronically excited state is described as a *quencher* and is said to *quench* that state. Thus, if the original excited state is luminescent, *quenching* will be observed as a diminution of the intensity (quantum yield) of light emission. The process can be represented as:

$$M^* \xrightarrow{Q} M' \qquad (4.1)$$

(where M′ is the ground state or another excited state of M, and Q is the quencher).

It must be emphasized that a reduction in (for example) fluorescence is *prima facie* evidence of electronic quenching only if a Boltzmann distribution over vibrational levels of the emitting state is attained before emission occurs. In the gas phase, the frequently observed variation of ϕ_f with pressure is found often to be due to collisionally-induced vibrational relaxation rather than electronic quenching.

The quenching process of (4.1) is of such generality that, as might be expected, it occurs by many different mechanisms and is induced by many different substances. Of these, oxygen is the most ubiquitous and one of the most efficient in that each encounter with an excited molecule leads to quenching. For this reason it is essential in all quantitative work to reduce the concentration of dissolved oxygen to the smallest possible value, either by bubbling oxygen-free nitrogen through the solution or, better, by degassing with several 'freeze–pump–thaw' cycles.† For similar reasons rigorous standards of purity are essential in all work on luminescence. Solvents should be non-fluorescent, and substrates should be purified by chromatography, zone-refining etc. until τ_f is constant.

Quenching processes, with the exception of certain kinds of electronic energy transfer, seem to be collisional and therefore subject to the Wigner

† The vessel is cooled in liquid nitrogen, evacuated to pump off the supernatant gas, sealed from the pump and allowed to thaw to release dissolved gases.

spin-conservation rules.† The bimolecular entity in which the quenching occurs can be either an encounter complex or an excimer/exciplex.‡ The distinction between these two sorts of entities seems to be this, that in an encounter complex, represented in this book as (M*... Q), the components are separated by distances of the order of 7 Å or more and have random relative orientations, the only requirement being some 'significant' overlap of the orbitals of the components. On the other hand, excimers and exciplexes [represented by (MM)* and (MQ)*] are entities occupying energy minima in the excited state potential surface and therefore having definite geometries. In aromatic excimers the components are organized into parallel planes (see below).

In principle, quenching can occur in either encounter complexes or excimers/exciplexes or both. The actual entity implicated has been defined for only relatively few systems. Seen in these terms, bimolecular quenching depends on the reversible formation of an encounter complex or exciplex which subsequently relaxes to ground state entities by various modes. Figure 4.1 depicts relaxation modes open to exciplexes. Similar diagrams may be constructed for encounter complexes or excimers.

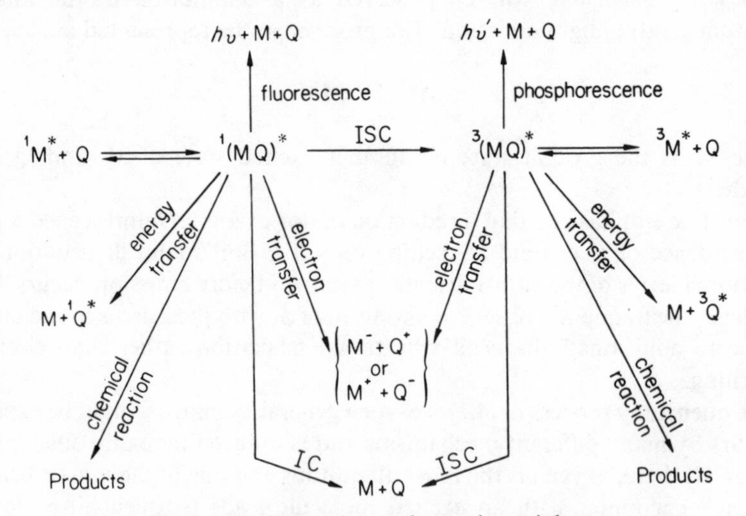

Figure 4.1. Relaxation pathways in exciplexes

† If two reactants A and B with spin S_A and S_B react on collision to give products X and Y with spins S_X and S_Y, then the spin of the transition state or collision complex can only have one of the following values:

$$(S_A + S_B), (S_A + S_B - 1), (S_A + S_B - 2) \ldots |S_A - S_B|$$

Similarly, the transition state giving rise to the products must have one of the spin values:

$$(S_X + S_Y), (S_X + S_Y - 1), (S_X + S_Y - 2) \ldots |S_X - S_Y|$$

Therefore for a collisional process to be spin-allowed, the two sequences must have a number in common.

‡ An exciplex is the excited state of a complex and is formed by the association of two different species, one excited and the other in its ground state. An excimer is an *excited dimer*, i.e. an exciplex involving the ground and excited states of the same species. These entities are discussed in the next section.

Quenching by electronic energy transfer and by the routes indicated in Figure 4.1 form the main subject matter of this chapter, but an account of the nature of excimers and exciplexes and of the kinetics of quenching is required first.

4.2 EXCIMERS[1]

It is commonly observed that an increase in the concentration of a solute is accompanied by a decrease in the intensity (quantum yield) of its fluorescence. This phenomenon is called *concentration quenching* and has been known for some time. More recently it has been found that such quenching is often accompanied by the appearance of a new emission at longer wavelengths, the intensity of which increases with concentration. For example, the violet fluorescence of pyrene (4.2) in dilute solution is gradually replaced by a blue fluorescence with increasing pyrene concentration (Figure 4.2). Förster and Kasper[2] showed that

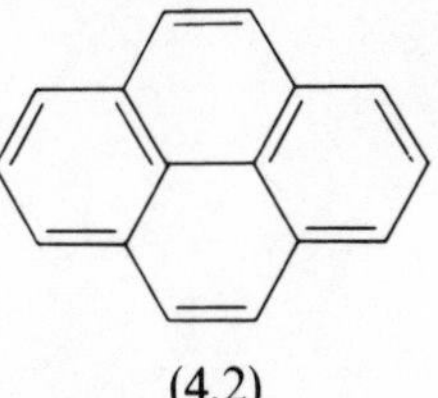

(4.2)

these phenomena could be explained by the formation and fluorescence of a pyrene excimer (4.3).

$$\mathrm{M} + h\nu \underset{\text{monomer fluorescence}}{\rightleftarrows} {}^1\mathrm{M}^* \overset{\mathrm{M}}{\rightleftarrows} \underset{\text{excimer}}{{}^1(\mathrm{MM})^*} \xrightarrow[\text{fluorescence}]{\text{excimer}} \mathrm{M} + \mathrm{M} + h\nu' \qquad (4.3)$$

It has since been found that excimer formation is widespread among aromatic hydrocarbons, though with simpler systems such as benzene, naphthalene and their methyl derivatives lower temperatures and higher concentrations than those necessary for pyrene are required.

Such complexes exist only in the excited state, being dissociated and therefore undetectable in the ground state. Figure 4.3 illustrates the points that emission from an excimer will be devoid of fine structure and will occur at longer wavelengths than that of its components.

Figure 4.4 shows the time-dependence of emission from pyrene monomer and excimer after flash excitation. Kinetic analysis[3] of such decay curves permits evaluation of the lifetime of the excimer and of the rate constants for its formation and dissociation. The rate constants for excimer formation (also obtainable from measurements of the quantum yield of emission as a function of concentration) are frequently found to be close to the diffusion-controlled rate constant k_{diff}, except where substituents can exercise a steric effect.

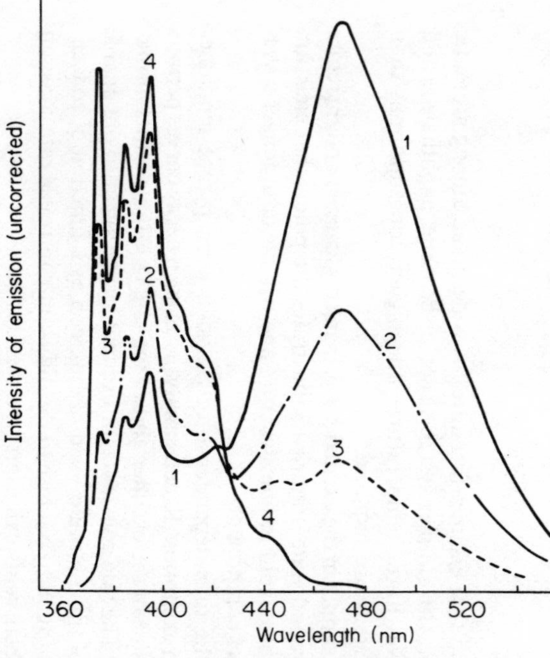

Figure 4.2. Normal fluorescence of pyrene in ethanol: (1) 3×10^{-3} M, (2) 10^{3} M, (3) 3×10^{-4} M, (4) 2×10^{-6} M. The instrumental sensitivity settings for curves 1 and 4 were approximately 0·6 and 3·7 times that for curves 2 and 3. The short wavelength ends of the spectra in the more concentrated solutions are distorted by self-absorption. (From C. A. Parker and C. G. Hatchard, *Trans. Faraday Soc.*, **59**, 284 (1963), reproduced by permission of the Chemical Society)

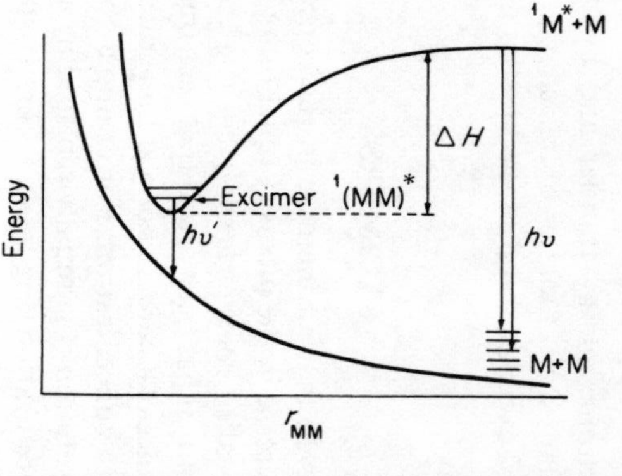

Figure 4.3. Schematic energy surfaces showing excimer formation and emission. The emission to the nonquantized ground state is structureless

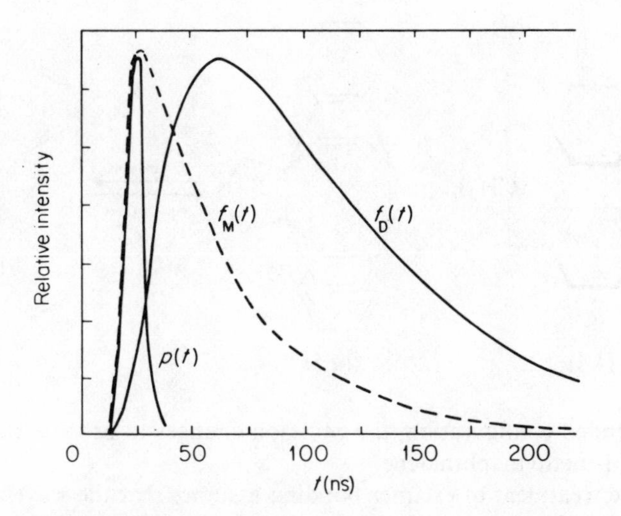

Figure 4.4. Pyrene (5×10^{-3} M) in cyclohexane. Monomer fluorescence response $f_M(t)$ and excimer fluorescence response $f_D(t)$ to excitation light pulse $p(t)$ (from J. B. Birks, D. J. Dyson and I. H. Munro, *Proc. Roy. Soc. A*, **275**, 575 (1963), reproduced by permission of the Royal Society)

From an examination of the temperature-dependence of the emission and by other methods it is possible to evaluate thermodynamic parameters. For pyrene[4] the enthalpy of dissociation (ΔH in Figure 4.3) is ~ 40 kJ mol^{-1} (10 kcal mol^{-1}), and the entropy of dissociation is ~ 80 J K^{-1} (20 cal K^{-1}), showing that this excimer has a strongly bonded and rigid structure. Excimers derived from other hydrocarbons have smaller enthalpies (~ 20 kJ mol^{-1}, 5 kcal mol^{-1}), though the dissociation entropies are usually similar.

4.2.1 Excimer Structure and Bonding

Excimers derived from aromatic hydrocarbons seem to adopt a sandwich structure with the molecular planes separated by $\sim 3{\cdot}3$ Å. Evidence supporting this point is the fact that those crystals of hydrocarbons in which the molecules are stacked in parallel planes have a fluorescence emission closely resembling that of the corresponding excimer,[5] and that in the paracyclophane series the 4,4′-compound (4.4, $n = 4$) with an interplanar distance of 3·73 Å is the lowest member of the series to exhibit an alkylbenzene absorption spectrum and a broad structureless intramolecular excimer emission. With the 4,5′- and 6,6′-paracyclophanes the emission returns to that of an alkyl benzene.[6]

Furthermore, photolysis of (4.5) in a low temperature glass generates (4.6) in the unstable sandwich geometry, in which form it has a fluorescent emission similar to the excimer emission of naphthalenes in fluid solution. When the glass is thawed and then refrozen, permitting the sandwich system to attain the more

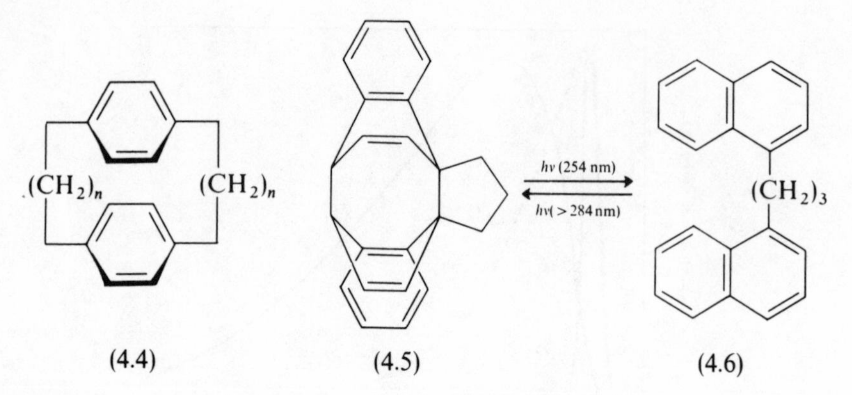

stable extended configuration, the emission changes to become closely similar to that of 1-methylnaphthalene.

A simple treatment of excimer bonding assumes that the wavefunction is of the form (4.7).

$$\psi_{\text{excimer}} = a\psi_{\text{MM*}} + b\psi_{\text{M*M}} + c\psi_{\text{M}^-\text{M}^+} + d\psi_{\text{M}^-\text{M}^+} \qquad (4.7)$$

This implies both exciton resonance [MM* ↔ M*M] and charge–transfer resonance [$M^-M^+ \leftrightarrow M^+M^-$]. Calculations[7] show that agreement between theoretical and observed values of excimer fluorescence can only be obtained if the components adopt a sandwich configuration with an interplanar distance of 3·0–3·6 Å, and if both exciton and charge–transfer resonance are invoked.

Excimer formation also occurs with non-aromatic molecules, and although the discussion has been directed to singlet excimers, the observation of excimer phosphorescence establishes the existence of triplet excimers also.

4.3 EXCIPLEXES[1]

Phenomena similar to those described for excimers are observed in solutions of mixed solutes. For example,[8] addition of diethylaniline to a solution of anthracene in toluene quenches the fluorescence of the latter and replaces it by a new structureless emission at longer wavelengths (Figure 4.5). This is ascribed to the formation of an exciplex (*exci*ted com*plex*) ^{1}(anthracene-diethylaniline)*. The term exciplex refers to an excited complex of definite stoichiometry (usually 1 : 1) formed between an excited species and one or more different molecules in their ground states (4.8).

$$\text{M} \xrightarrow{h\nu} \text{M*} \xrightarrow{\text{Q}} \underset{\text{exciplex}}{(\text{MQ})^*} \qquad (4.8)$$

Exciplexes are polar entities. The dipole moments of exciplexes from aromatic hydrocarbons and aromatic tertiary amines, estimated from the red shift of the emission peak with increasing dielectric constant of solvent, have values of >10 debyes.[9] Excimers, on the other hand, have zero dipole moments.

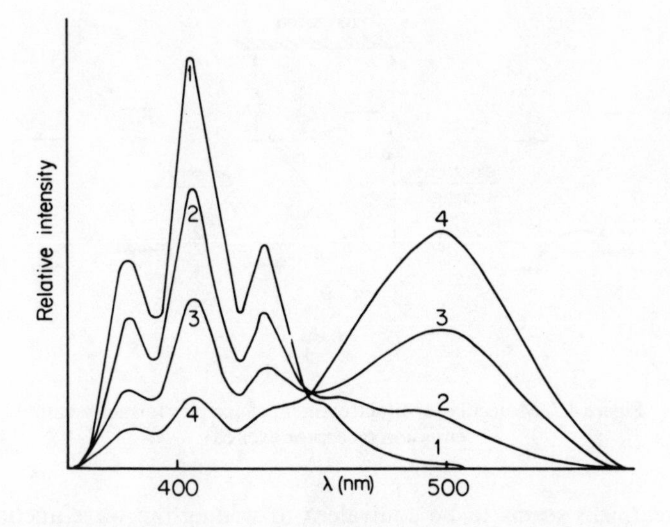

Figure 4.5. Fluorescence spectra of anthracene (3×10^{-4} M) in toluene in presence of diethylaniline at concentrations 0·000 M (1); 0·005 M (2); 0·025 M (3); 0·100 M (4). (From A. Weller, in S. Claesson (ed.), *Fast Reactions and Primary Processes in Chemical Kinetics*, (1967), John Wiley and Sons Ltd.)

A simple molecular orbital treatment due to Weller[10] provides an explanation of the large dipole moment and other properties of exciplexes. It assumes that in an exciplex an electron is transferred from one component, D (donor) to the other, A (acceptor). Figures 4.6 and 4.7 illustrate the situation when the donor and the acceptor respectively are excited.

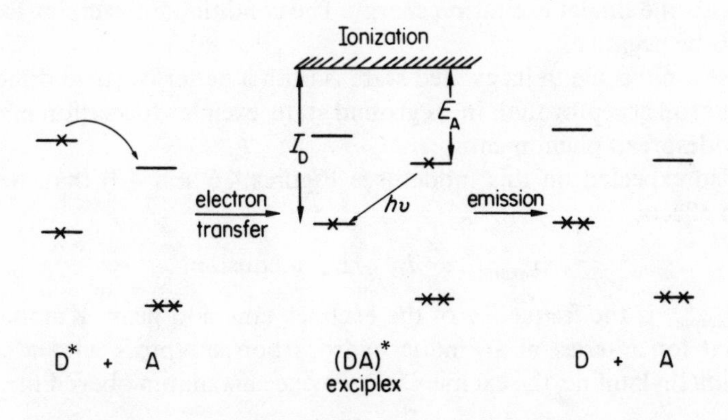

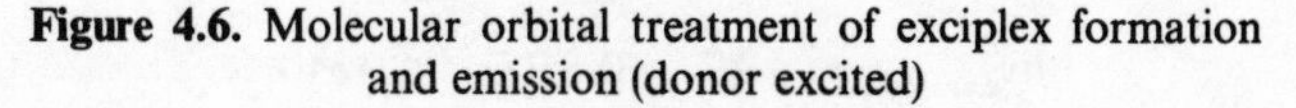

Figure 4.6. Molecular orbital treatment of exciplex formation and emission (donor excited)

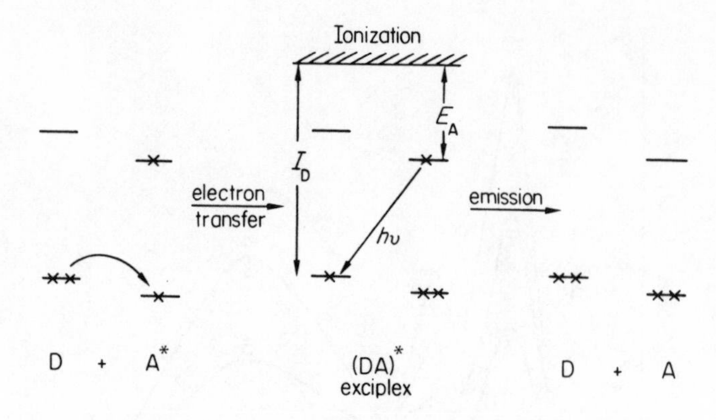

Figure 4.7. Molecular orbital treatment of exciplex formation and emission (acceptor excited)

This treatment seems to be equivalent to writing the wavefunction of the exciplex as (4.9) and neglecting all terms except the fourth.

$$\psi_{\text{exciplex}} = a\psi_{\text{D}^*\text{A}} + b\psi_{\text{DA}^*} + c\psi_{\text{D}^-\text{A}^+} + d\psi_{\text{D}^+\text{A}^-} \tag{4.9}$$

One can readily formulate the conditions for the formation of exciplexes on the charge–transfer model. Since the ionization potential (I) and electron affinity (E) are measures of the energies of the highest bonding and lowest anti-bonding molecular orbitals respectively, ΔG, the energy of formation of the separate ions A^- and D^+ from A and D in their ground states, is $I_D - E_A$. Since we actually start with, say, A*, the energy of formation is diminished by the excitation energy possessed by A*. Bringing the separated ions A^- and D^+ to their equilibrium distance (r) in the exciplex reduces ΔG by an electrostatic term so that we can write

$$\Delta G = I_D - E_A - E_{00} - e^2/r \tag{4.10}$$

where E_{00} is the singlet excitation energy. The condition for exciplex formation is that ΔG be negative.

Because a molecule in its excited state is both a better electron donor and a better electron acceptor than in its ground state, exciplex formation is expected to be a widespread phenomenon.

It is also expected on this model (see Figures 4.6 and 4.7) that, neglecting solvation effects,

$$h\nu_{\text{exciplex}} = I_D - E_A + \text{constant} \tag{4.11}$$

where ν_{exciplex} is the frequency of the exciplex emission peak. Knibbe *et al.*[11] found that for a series of aromatic hydrocarbon acceptors with a common donor, dimethylaniline, the exciplex fluorescence maximum obeyed the relation (4.12).

$$h\nu_{\text{exciplex}} = 1{\cdot}17 - 0{\cdot}65(E_{\text{A}^-/\text{A}})\,(\text{eV}) \tag{4.12}$$

Since the reduction potential $E_{A^-/A}$ is linearly related to the electron affinity E_A of the acceptor, these results support the model and clearly demonstrate the charge–transfer character of the exciplex.

It should be recognized that exciplex formation is not restricted to aromatic systems and that there is no requirement that exciplexes should necessarily luminesce. If the sum of the rate constants for radiationless processes involving the exciplex (see Figure 4.1) is sufficiently high, the lifetime of the exciplex will be so short that its emission may be indetectable. Similarly, very short-lived non-luminescent exciplexes may be generated if their binding energy is $< kT$ (where k is Boltzmann's constant). Since exciplexes/excimers do not exist in the ground state, direct evidence for their formation can only be obtained from their emission, or perhaps from their absorption characteristics following their generation by a high-intensity light flash.

Exciplexes or excimers, though difficult to detect, seem to be implicated in many photochemical processes, e.g.

(i) Photodimerization of polyacenes[12] (4.13).

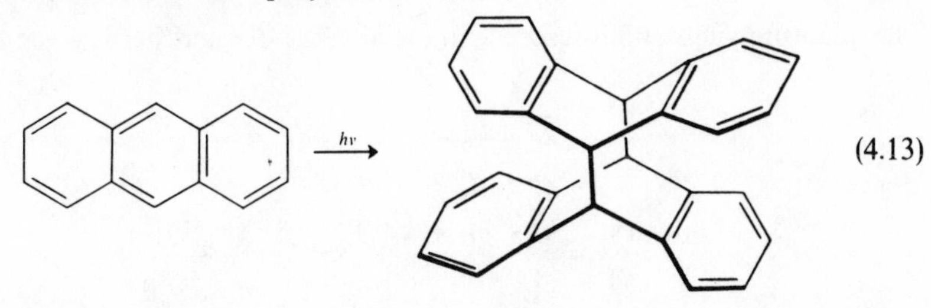

(4.13)

(ii) Photoaddition of ketones to electron-deficient olefins[13] (4.14).

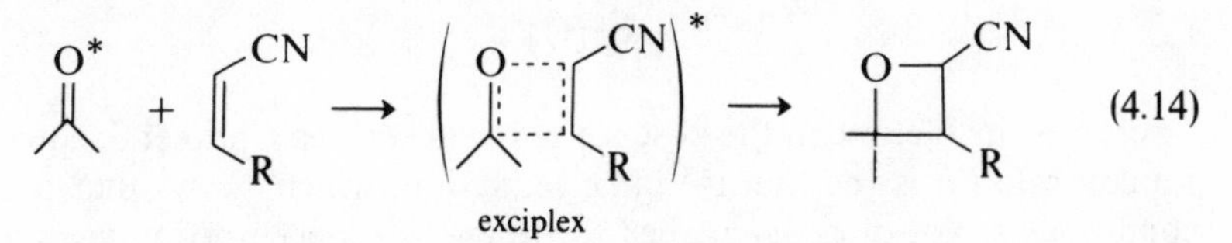

(4.14)

(iii) Triplet–triplet annihilation (4.15, see Chapter 3, p. 83).

$$^3M^* + {}^3M^* \rightarrow {}^1(MM)^* \rightarrow {}^1M + {}^1M^* \tag{4.15}$$

(iv) In scintillation counting, an ionizing particle is stopped by, and excites, molecules of an aromatic solvent. These transfer their energy to, and excite, a fluorescent solute or scintillator, the emission of which is detected by a photomultiplier tube. The required energy transfer from the site of initial excitation to the scintillator could occur by non-radiative energy transfer (see Chapter 4, p. 121), or alternatively by the reversible formation of excimers derived from the solvent (4.16).

$$^1M^* + {}^1M \rightarrow {}^1(MM)^* \rightarrow {}^1M + {}^1M^* \tag{4.16}$$

(v) Exciplexes are also probably involved in many quenching processes (see below).

4.4 THE KINETICS OF QUENCHING

Because kinetics are basic to any real understanding of quenching phenomena, a simple treatment is now given (the more complex aspects are deferred until Chapter 5, section 5.2.2). Consider the situation of (4.17), where an excited species fluoresces, decays non-radiatively or is quenched, with rate constants k_f, k_D and k_q respectively.

$$\mathrm{M} \xrightarrow[I]{h\nu} \mathrm{M}^* \begin{cases} \xrightarrow{k_f} \mathrm{M} + h\nu \\ \xrightarrow{k_D} \mathrm{M} \\ \xrightarrow[\mathrm{Q}]{k_q[\mathrm{Q}]} \mathrm{M} + \mathrm{Q} \end{cases} \tag{4.17}$$

Applying the steady-state approximation to the concentration of M* gives:

$$I = [\mathrm{M}^*](k_f + k_D + k_q[\mathrm{Q}])$$

$$\therefore \quad \phi_f = \frac{k_f[\mathrm{M}^*]}{I} = \frac{k_f}{k_f + k_D + k_q[\mathrm{Q}]}$$

The quantum yield of fluorescence in the absence of quencher is given by:

$$\phi_f^0 = \frac{k_f}{k_f + k_D}$$

Hence

$$\frac{\phi_f^0}{\phi_f} = \frac{k_f + k_D + k_q[\mathrm{Q}]}{k_f + k_D} = 1 + \frac{k_q[\mathrm{Q}]}{k_f + k_D}$$

and

$$\frac{\phi_f^0}{\phi_f} = 1 + k_q\tau[\mathrm{Q}] \tag{4.18}$$

where τ is the lifetime in the absence of quencher (see Chapter 3, equation 3.1). Equation (4.18) is the Stern–Volmer equation, which shows that under ideal conditions a plot of ϕ_f^0/ϕ_f against the quencher concentration gives a straight line of gradient $k_q\tau$. Hence if τ is known, then the quenching rate constant is immediately obtained.

For many systems, k_q in mobile solvents is of the order of 10^9–10^{10} l mol^{-1} s^{-1}, i.e. close to the diffusion-controlled rate constant k_{diff}. This suggests that in these cases quenching is so rapid that the rate-determining step is the diffusion of the quencher to within the 'active sphere' of M*. The position is best examined quantitatively. Suppose that bimolecular quenching cannot occur until Q and M* have formed an encounter complex/exciplex (4.19).

$$\mathrm{M}^* + \mathrm{Q} \underset{k_{-1}}{\overset{k_{diff}}{\rightleftarrows}} (\mathrm{M}^* \cdots \mathrm{Q}) \xrightarrow{k_c} \mathrm{M} + \mathrm{Q} \tag{4.19}$$

Here, k_c is the actual quenching rate constant within the complex. Then, under steady-state conditions,

$$k_{diff}[\mathrm{M}^*][\mathrm{Q}] = [(\mathrm{M}^* \cdots \mathrm{Q})](k_c + k_{-1})$$

and

$$\text{rate of quenching} = k_c[(M^* \cdots Q)]$$

$$= \frac{k_c k_{diff}}{k_c + k_{-1}}[M^*][Q]$$

$$\therefore \quad k_q\,(\text{observed}) = \frac{k_c k_{diff}}{k_c + k_{-1}} \tag{4.20}$$

There are three cases of interest:

(a) If $k_c \gg k_{-1}$, then $k_q\,(\text{obs}) \sim k_{diff}$ and will be dependent upon solvent viscosity;

(b) If $k_c \ll k_{-1}$, then k_q (obs) $\sim k_c \,.\, k_{diff}/k_{-1} = k_c K$ where K is the equilibrium constant for the formation of the complex. This means that with weak quenchers k_q (obs) will be independent of solvent viscosity, as are the rate constants of ordinary bimolecular ground state reactions;

(c) If k_c and k_{-1} are of the same order, then k_q (obs) will be less than k_{diff}.

Because diffusion dominates bimolecular quenching processes in solution, it is important to have a reliable method of estimating k_{diff}. One commonly used relation is the Debye equation (4.21).

$$k_{diff} = \frac{8RT}{3000\eta}(\text{l mol}^{-1}\,\text{s}^{-1}) \tag{4.21}$$

R is the gas content, T is the temperature, and η is the viscosity (in poise). Although the derivation of this equation depends upon a number of simplifying assumptions (e.g., that the diffusing species are spherical, of the same diameter, with the same interaction radii, and that the microscopic and macroscopic viscosities of the solvent are the same), it succeeds reasonably well in predicting the maximum values of k_q in a variety of solvents. However, critical tests of its validity reveal deficiencies. For example, k_q for quenching the fluorescence of a number of polynuclear aromatic hydrocarbons by oxygen is found[14] to be several times as large as k_{diff} calculated from equation (4.21). It seems that this equation fails when the solute molecules differ considerably in size or when they are small by comparison with solvent molecules. Hence, for the quenching of naphthalene phosphorescence by 1-iodonaphthalene the ratio k_q/k_{diff} rises to 4·5 in liquid paraffin.[15] Even for molecules of comparable size in normal solvents, equation (4.21) tends to underestimate k_{diff}, and the modified relation (4.22) seems to be more satisfactory.[15]

$$k_{diff} = \frac{8RT}{2000\eta} \tag{4.22}$$

4.5 QUENCHING PROCESSES AND QUENCHING MECHANISMS

There are many quenching processes, and Figure 4.8 illustrates a possible classification scheme.

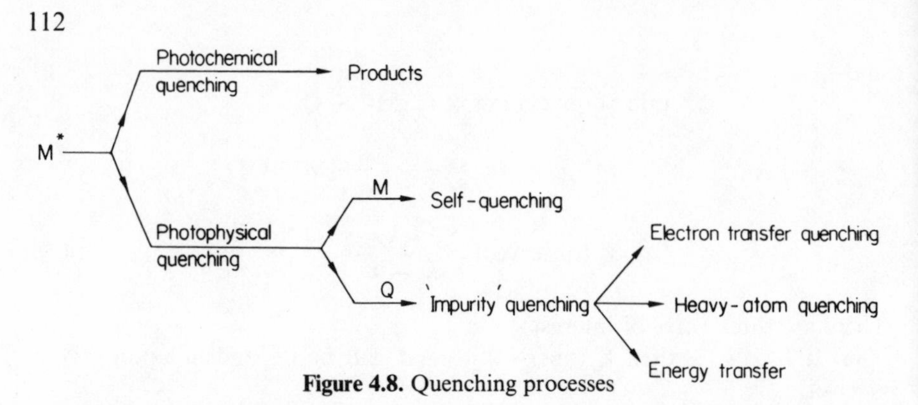

Figure 4.8. Quenching processes

Quenching by photochemical reaction forms the subject matter of organic photochemistry and will not be discussed here. Photophysical quenching, which does not lead to new ground state products, can be divided into self-quenching (see excimers, section 4.2), in which the quenching species is M, and 'impurity' quenching, where the quencher is some other chemical species. This last category can be further split into quenching by electron transfer, by heavy-atom effects and by energy transfer, and these three processes will now be analysed.

4.5.1 Electron Transfer Quenching

The emission by an exciplex of an aromatic hydrocarbon with an amine in solvents of increasing polarity is observed not only to shift to the red but also to diminish in intensity, becoming zero in solvents with high dielectric constant such as acetonitrile.[10] Flash photolysis experiments in acetonitrile show the transient generation of the hydrocarbon radical anion and the amine radical cation, and this implies the total transfer of an electron from donor to acceptor. Similarly the products derived from the irradiation of naphthalene with triethylamine in acetonitrile indicate[16] that electron transfer is a primary process (4.23).

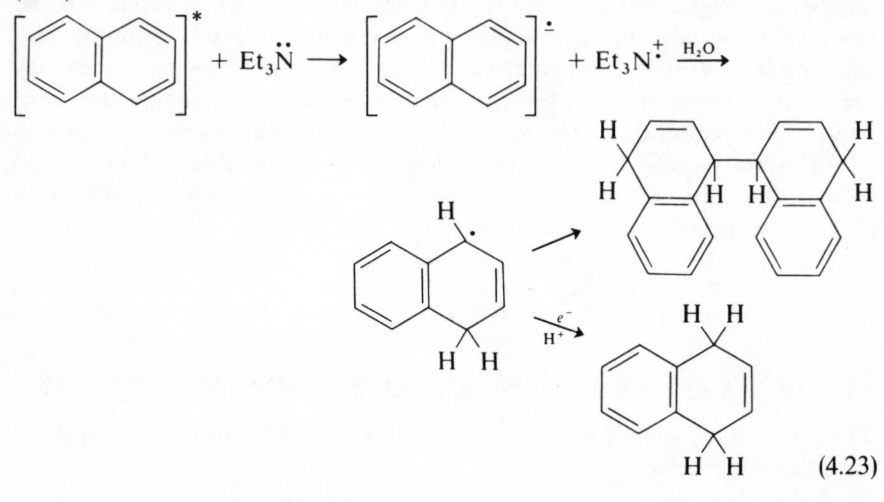

(4.23)

The simplest hypothesis is that in polar solvents the exciplex dissociates into solvated ions (4.24).

$$^1(\mathrm{AD})^* \rightarrow \mathrm{A}_s^{\dot{-}} + \mathrm{D}_s^{\dot{+}} \tag{4.24}$$

However, it has been shown that, for the pyrene–dimethylaniline system, increasing the dielectric constant of the solvent reduces the lifetime of the exciplex very much less than it reduces the intensity of exciplex fluorescence. These facts are explained by postulating a competing electron transfer process occurring in an encounter complex between $^1\mathrm{A}^*$ and D, giving rise to solvated ion pairs (4.25).

$$^1\mathrm{A}^* + \mathrm{D} \rightarrow {^1(\mathrm{A}^{\overset{*}{-}}\text{-}\mathrm{D})}$$

encounter complex $\xrightarrow{k_c}$ $^1(\mathrm{AD})^*$ (exciplex); encounter complex $\xrightarrow{k_q}$ $\mathrm{A}_s^{\dot{-}} + \mathrm{D}_s^{\dot{+}}$ (solvated ion pair)

$^1(\mathrm{AD})^*$ $\xrightarrow{k_f}$ $h\nu + \mathrm{A} + \mathrm{D}$; $^1(\mathrm{AD})^*$ $\xrightarrow{k_d}$ $\mathrm{A} + \mathrm{D}$; $^1(\mathrm{AD})^*$ $\xrightarrow{k_r}$ $\mathrm{A}_s^{\dot{-}} + \mathrm{D}_s^{\dot{+}}$ (4.25)

k_q and k_r will both increase with solvent polarity, but k_c will be independent of this parameter. Thus the exciplex lifetime, $\tau_e = (k_f + k_d + k_r)^{-1}$ will decrease in polar solvents. However, the quantum yield for exciplex fluorescence, which equals the probability of exciplex formation multiplied by $k_f\tau_e$, is given by (4.26).

$$\phi_f = \frac{k_c}{k_c + k_q} \,.\, k_f\tau_e \tag{4.26}$$

Hence ϕ_f, the product of two solvent-dependent terms, will decrease with solvent polarity more rapidly than will τ_e.

The distance between the partners in an encounter complex is estimated to be ~7 Å or more, considerably larger than that obtaining in excimers/exciplexes, and this implies an ill-defined mutual orientation of the components and perhaps requires little more than orbital overlap somewhere in the complex. Whether such entities are to be regarded as exciplexes is a matter of definition.

Further evidence relating to the role of encounter complexes in electron transfer quenching was obtained by Rehm and Weller,[17] who proposed the scheme (4.27) to account for the kinetics of the quenching of a number of polynuclear aromatics by electron donors in acetonitrile.

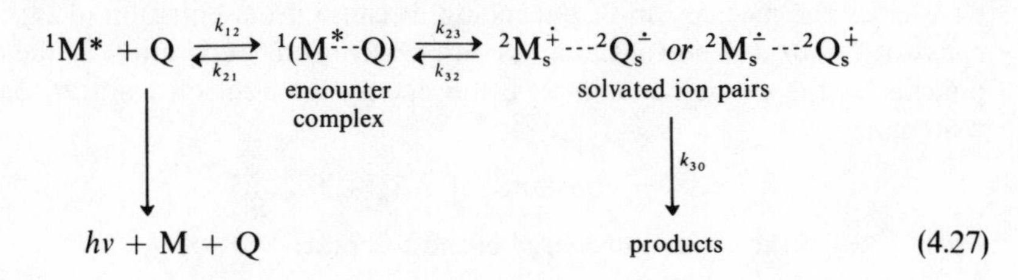

(4.27)

Given the linear relationship between ionization potentials, electron affinities and polarographic redox potentials, the free energy of the electron transfer process can be seen from Figure 4.6 to be:

$$\Delta G_{23} = E_{(D/D^+)} - E_{(A^-/A)} - \frac{e^2}{\varepsilon a} - \Delta E_{00} \tag{4.28}$$

where the first two terms are redox potentials, the third term is the Coulomb energy released in bringing the ions to within the encounter distance a, ε is the dielectric constant of the medium, and ΔE_{00} is the electronic excitation energy. A kinetic analysis[17] of scheme (4.27) gives the quenching rate constant as:

$$k_q = \frac{2\cdot0 \times 10^{10}}{1 + 0\cdot25[\exp(\Delta G_{23})/RT + \exp(\Delta G^{\ddagger}_{23})/RT]}(\mathrm{l\ mol^{-1}\ s^{-1}}) \tag{4.29}$$

where $\Delta G^{\ddagger}_{23}$, which can be calculated from ΔG_{23}, is the energy of activation for the formation of the solvated ion pairs from the encounter complex. Thus k_q can be calculated from spectroscopic and electrochemical data. The experimental data shown in Figure 4.9 conform to a relationship of the type in equation (4.29), since $k_q \approx k_{diff}$ when ΔG_{23} is more exothermic than 20–40 kJ mol^{-1} (5–10 kcal mol^{-1}), then k_q falls as ΔG_{23} becomes less exothermic, and finally k_q becomes proportional to $\exp(-\Delta G_{23}/RT)$ when ΔG_{23} becomes endothermic by more than 20 kJ mol^{-1} (5 kcal mol^{-1}). The results are incompatible with electron transfer quenching within an exciplex.

The quenching of the singlet states of aromatic hydrocarbons by dienes and by quadricyclane (4.30) and of azo-compounds (e.g. 4.31) by dienes and olefins

has been shown not to be due to energy transfer nor to chemical quenching. In these systems, k_q is rather insensitive to solvent polarity, but is affected by substitution which inhibits close approach of the quencher. The quenching seems to be due to exciplex formation, even though exciplex emission is not observed. For example, the quenching of aromatic hydrocarbons by quadricyclane gives the results[18] shown in Figure 4.10. This clearly establishes electron transfer as the mechanism of quenching, because, from equation (4.28), for a constant donor and solvent and assuming a constant separation of the components of the encounter complex, the energy for electron transfer may be written as

$$\Delta G = \text{constant} - [{}^1\Delta E_A + E_{(A^-/A)}] \tag{4.32}$$

where ${}^1\Delta E_A$ is the excitation energy of the acceptor.

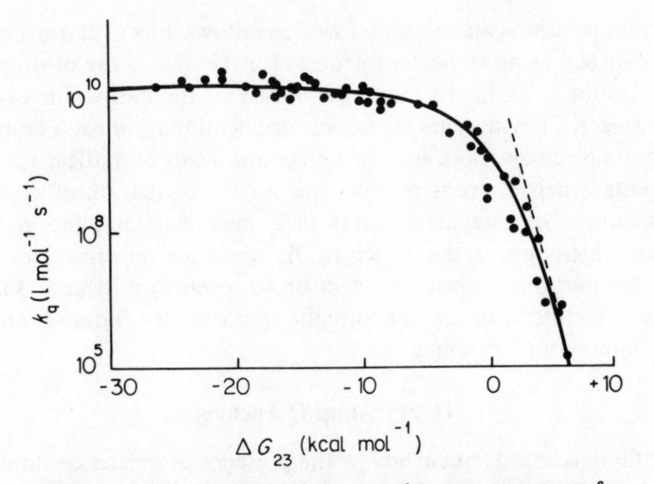

Figure 4.9. Relationship between the rate constant for quenching (k_q) of several polynuclear aromatics by electron donors, and the energy for the electron transfer process (ΔG_{23}). (From D. Rehm and A. Weller, *Ber. Bunsenges. Phys. Chem.*, **73**, 836 (1969), reproduced by permission of Verlag Chemie Gmbh.)

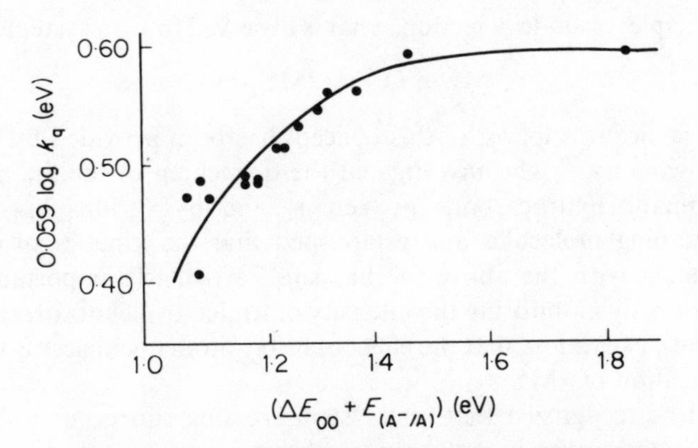

Figure 4.10. Correlation between rate constants for fluorescence quenching (k_q) by quadricyclane and electron affinities of the excited molecules (from B. S. Solomon, C. Steel and A. Weller, *Chem. Commun.*, 927 (1969), reproduced by permission of the Chemical Society)

Similarly, Evans[19] established for the quenching of the fluorescence of azo-compound (4.31) by dienes, and of naphthalene by dienes and olefins, a linear relationship between the ionization potential of the quencher and $\ln[k_q/(k_{diff} - k_q)]$, where k_q is the Stern–Volmer quenching rate constant, and further showed that such a dependence was to be expected if electron transfer were the rate-limiting step.

The current position can be summarized as follows. Electron transfer has been clearly established as an important process for the quenching of singlet excited states, but doubt exists as to the exact nature of the entities involved in the electron transfer. They may be exciplexes or encounter/collision complexes. In polar solvents electron transfer is complete and leads to radical ions. In non-polar solvents, where there is no evidence for the formation of products as a result of the quenching step, electron transfer may be incomplete, giving rise to an exciplex which may relax either by fluorescence or radiationlessly by a 'return' of the partially transferred electron to regenerate M and Q in their S_0 or T_1 states. The last cases may be thought of as exciplex-induced internal conversion or intersystem crossing.

4.5.2 Heavy Atom Quenching

Molecular fluorescence is quenched by the presence of species containing heavy atoms, and it seems that the phenomenon is due to the formation of a singlet exciplex (or encounter complex) which, because of the heavy atom effect, undergoes enhanced intersystem crossing to the triplet exciplex, followed by dissociation into its components.

$$^1M^* + Q \rightarrow {}^1(MQ)^* \xrightarrow{isc} {}^3(MQ)^* \rightarrow {}^3M^* + Q$$

Since the exciplexes elude detection, what is observed in such systems is:

$$^1M^* + Q \rightarrow {}^3M^* + Q$$

Much evidence in support of this concept has been provided by Wilkinson and his co-workers,[20] who investigated the quenching of the fluorescence of several aromatic hydrocarbons by xenon† and by various bromine- and iodine-containing molecules and established that the kinetics of quenching were consistent with the above mechanism. Particularly important was the demonstration, by monitoring the intensity of triplet–triplet absorption of $^3M^*$ following flash excitation, that the effect of heavy atom quenchers is to increase the concentration of $^3M^*$.

It should be recognized that intersystem crossing can occur in exciplexes, as in other systems, even in the absence of heavy atoms.

4.5.3 Quenching by Oxygen[21] and Paramagnetic Species[22]

The quenching of the excited states of many organic molecules by oxygen is diffusion-controlled. Quenching of singlet states seems to occur both by collisional energy transfer (4.33), generating singlet oxygen,

$$^1M^* + {}^3O_2 \rightarrow {}^3M^* + {}^1O_2({}^1\Delta_g) \tag{4.33}$$

and by the *spin-allowed* catalysed intersystem crossing (4.34).

† Xenon is an ideal heavy atom quencher, since it is chemically inert, optically transparent and readily soluble in organic solvents.

$$^1M^* + {}^3O_2 \rightarrow {}^3M^* + {}^3O_2 \quad (4.34)$$

This may be thought of as occurring *via* the sequence (4.35).

$$^1M^* + {}^3O_2 \xrightarrow{k_{diff}} {}^3(MO_2)_n^* \xrightarrow{ic} {}^3(MO_2)_1^* \xrightarrow{\text{spin-allowed}} {}^3M^* + {}^3O_2 \quad (4.35)$$

highly excited triplet exciplex; lowest excited triplet exciplex

The quenching of triplet states, which could also occur by catalysed intersystem crossing to the ground state (4.36), actually seems to be dependent upon triplet energy transfer (see p. 125) generating singlet oxygen (4.37).

$$^3M^* + {}^3O_2 \rightarrow {}^3(MO_2)_n^* \rightarrow {}^3(MO_2)_1^* \xrightarrow{\text{spin-allowed}} {}^1M + {}^3O_2 \quad (4.36)$$

$$^3M^* + {}^3O_2 \xrightarrow{\text{spin-allowed}} {}^1M + {}^1O_2^* \quad (4.37)$$

The free radical nitric oxide (^2NO) is also a highly efficient quencher of excited singlet states. By analogy with oxygen, it probably quenches by enhancing the rate of intersystem crossing (4.38).

$$^1M^* + {}^2NO \rightarrow {}^2(M.NO)_n^* \rightarrow {}^2(M.NO)_1^* \xrightarrow{\text{spin-allowed}} {}^3M^* + {}^2NO \quad (4.38)$$

4.5.4 Electronic Energy Transfer[24]

In this phenomenon, which is of crucial importance in photochemistry, an excited donor molecule D* collapses to its ground state with the simultaneous transfer of its electronic excitation energy to an acceptor molecule A which is thereby promoted to an excited state (4.39).

$$D^* + A \rightarrow D + A^* \quad (4.39)$$

It should be noted that the acceptor can itself be an excited state, as in triplet–triplet annihilation (4.40, see Chapter 3, p. 83).

$$^3M^* + {}^3M^* \rightarrow {}^1M + {}^1M^* \quad (4.40)$$

What is observed in an energy transfer experiment is the quenching of the emission (or photochemistry) associated with D* and its replacement by the emission (or photochemistry) characteristic of A*. Hence, although the photons are absorbed by D, it is A which becomes excited. The processes resulting from A* generated in this manner are said to be *sensitized*. When the donor and acceptor are identical, the term energy migration is used, i.e. for:

$$M^* + M \rightarrow M + M^*$$

The principal mechanisms of electronic energy transfer are set out in Figure 4.11.

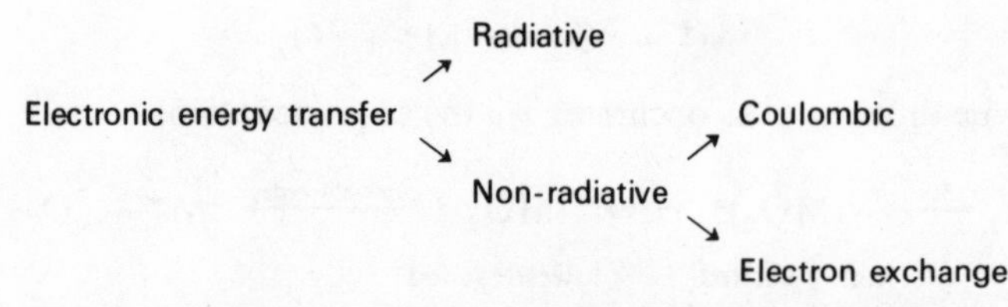

Figure 4.11. Principle mechanisms of electronic energy transfer

Although energy transfer can occur between other modes (translational, vibrational, rotational), the prime concern in this chapter will be with transfer of electronic energy. Radiative energy transfer depends on the capture, by the acceptor, of photons emitted by the donor (4.41).

$$\begin{aligned} D^* &\rightarrow D + h\nu \\ h\nu + A &\rightarrow A^* \end{aligned} \tag{4.41}$$

Clearly, such energy transfer can occur over immense distances: photochemical and photobiological changes occurring on Earth under the influence of sunlight are extreme examples of long-range radiative energy transfer. Non-radiative energy transfer induced by Coulombic interactions is also long-range in that it can occur over distances (50 Å or more) much larger than molecular dimensions. The electron exchange mechanism requires close approach (~10–15 Å), but not necessarily the contact, of the donor and acceptor species.

Spin conservation and other factors impose restrictions upon the relative multiplicities of the initial and final excited species involved in the electronic energy transfer, and the only common situations in organic photochemistry are those of (4.42).

$$\begin{array}{ll} \text{Singlet–singlet energy transfer:} & {}^1D^* + {}^1A \rightarrow {}^1D + {}^1A^* \\ \text{Triplet–singlet energy transfer:} & {}^3D^* + {}^1A \rightarrow {}^1D + {}^1A^* \\ \text{Triplet–triplet energy transfer:} & {}^3D^* + {}^1A \rightarrow {}^1D + {}^3A^* \\ \text{Oxygen quenching:} & {}^3D^* + {}^3O_2 \rightarrow {}^1D + {}^1O_2^* \end{array} \tag{4.42}$$

These phenomena will now be discussed in detail.

Radiative Energy Transfer

Since the process requires that the acceptor absorb photons emitted by the donor, it is obvious that the probability (rate) of transfer will depend upon (i) the quantum efficiency of emission by the donor θ_E, (ii) the number of acceptor molecules in the path of the emitted photon, (iii) the light-absorbing power of the acceptor, and (iv) the extent of the overlap between the emission spectrum of the donor and the absorption spectrum of the acceptor. This last requirement is expressed mathematically by the integral:

$$\int_0^\infty F_D(\bar{\nu})\varepsilon_A(\bar{\nu})\,d\bar{\nu} \tag{4.43}$$

where $F_D(\bar{\nu})$ is the emission spectrum of D*, and $\varepsilon_A(\bar{\nu})$ is the absorption spectrum of the acceptor A, both plotted on a wavenumber scale (see Figure 4.12).

The probability of radiative transfer, P_{rt}, in homogeneous solution can therefore be expressed as (4.44),

$$P_{rt} \propto \frac{[A]l}{\theta_E} \int_0^\infty F_D(\bar{\nu})\varepsilon_A(\bar{\nu})\,d\bar{\nu} \tag{4.44}$$

where l is the distance over which energy transfer occurs.

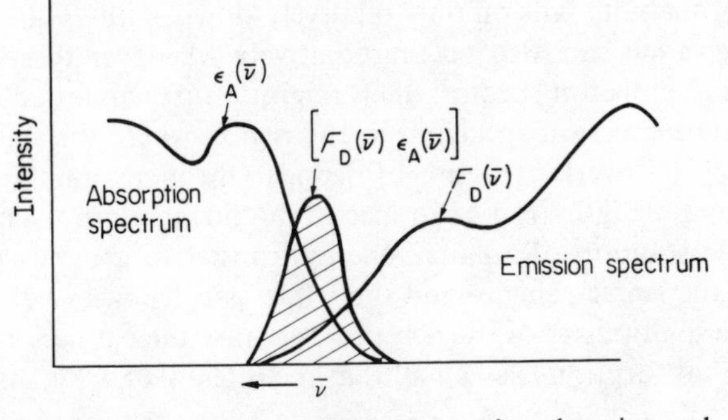

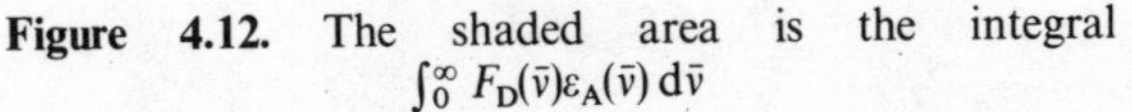
Figure 4.12. The shaded area is the integral $\int_0^\infty F_D(\bar{\nu})\varepsilon_A(\bar{\nu})\,d\bar{\nu}$

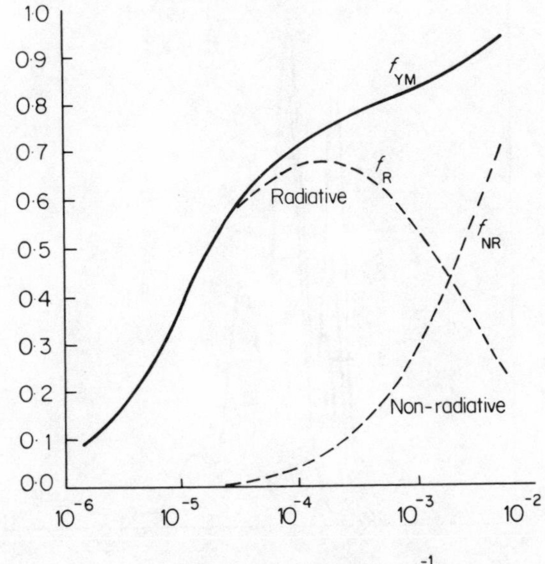

Figure 4.13. Singlet–singlet energy transfer from *p*-terphenyl (2·17 × 10^{-2} M) to tetraphenylbutadiene (TPB) in toluene solutions. Energy transfer quantum efficiency f_{YM} against TPB concentration. Radiative component, f_R; radiationless component, f_{NR}. (From J. B. Birks, *Photophysics of Aromatic Molecules*, (1970), John Wiley and Sons Ltd.)

The efficiency of such energy transfer depends also on the shape and size of the vessel used in the experiment, since photons emitted near the walls have less chance of capture by the acceptor than those emitted in the centre of the cell.

Radiative energy transfer, though frequently described as the 'trivial' mechanism because of its conceptual simplicity, is far from being such. It may be the dominant mechanism of energy transfer in dilute solutions (Figure 4.13), because its probability falls off only relatively slowly with dilution.

Energy migration can also occur radiatively whenever there is overlap of absorption and emission spectra. Such migration is manifested by changes in the fluorescence emission spectrum in the region where the fluorescence and absorption spectra overlap (Figure 4.14), due to the preferential reabsorption of light of these wavelengths. Emission spectra should therefore always be recorded on very dilute solutions. Radiative energy transfer to ground state species is restricted to the singlet–singlet and the triplet–singlet cases where there is no change in the multiplicity of the acceptor, because the extinction coefficient of singlet–triplet absorption is so small that the integral (4.43) is microscopic.

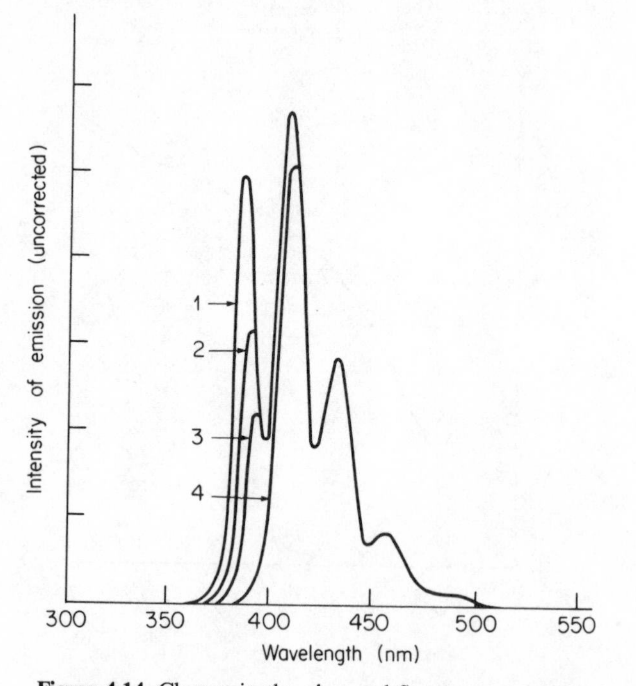

Figure 4.14. Change in the observed fluorescence spectra of anthracene in benzene solution at 20 °C due to reabsorption of the short wavelength band at higher concentrations (concentration increases 1 → 2 → 3 → 4). (From F. Wilkinson, in E. J. Bowen (ed.), *Luminescence in Chemistry*, (1968), reproduced by permission of Van Nostrand Reinhold Co. Ltd.)

Non-radiative Energy Transfer

If an acceptor molecule A has no interaction with a proximate excited donor molecule D*, then each, being 'unaware' of the other, will be in a state described by the appropriate Schrödinger equation. However, if there is an interaction between D* and A, however small, described by a perturbation Hamiltonian operator H', then the system (D* + A) is no longer a 'stationary' state of the total Hamiltonian ($H_{D*} + H_A + H'$). If there is an *isoenergetic* system (D + A*), in which the excitation now resides in the acceptor, then time-dependent perturbation theory (see Chapter 2) reveals that the effect of the interaction H' is to cause the system (D* + A) to evolve into the system (D + A*), and vice versa, with a probability given[25] by (4.45),

$$P \propto \rho \langle \psi_i | H' | \psi_f \rangle^2 \tag{4.45}$$

where ρ is the density of isoenergetic states, ψ_i describes the initial system (D* + A), and ψ_f describes the final system (D + A*).

In other words, the perturbation induces coupled radiationless transitions in both D and A so that excitation energy is shuttled back and forth between D and A and the rate of energy transfer (k_{ET}) is determined, according to equation (4.45), by the strength of the interaction H' between the components. If the coupling is sufficiently strong, as in some crystals, k_{ET} is so large that it is no longer possible to think of the excitation as being even temporarily localized on D or A, and it becomes associated with the system as a whole. This case, which requires that the absorption spectrum of the system (D + A) is not the sum of those of its components, will not be considered further.

We shall rather be concerned with the situation where H' is sufficiently small that $k_{ET} < k_{VR}$, the rate constant for vibrational relaxation. Under these circumstances, (i) D* will undergo energy transfer from its bottom vibrational level, (ii) A*, when formed, will promptly collapse to its bottom vibrational level, thereby destroying the degeneracy of (D* + A) and (D + A*) and making the energy transfer unidirectional, and (iii) k_{ET} will be given by the rate constant for forward transfer. This situation is pictorialized in Figure 4.15. The required degeneracy of the two systems (D* + A) and (D + A*) is readily achieved for organic molecules because of the availability of a multitude of vibrational and rotational sub-levels. Examination of Figure 4.15 also shows that the condition for irreversible energy transfer is that the $0 \rightarrow 0$ excitation energy of D (E_D) shall be significantly greater than that of A (E_A).

The state density factor ρ in equation (4.45) is a measure of the number of coupled isoenergetic donor and acceptor transitions. For organic molecules, which in solution exhibit broad spectra, ρ can be estimated from the overlap between the emission spectrum of the donor and the absorption spectrum of the acceptor, expressed in energy (i.e., wavenumber) units, or in other words by the integral (equation 4.43). It must be emphasized that although the rates of both radiative and non-radiative energy transfer depend upon this integral, the mechanisms are totally different. For the latter, energy transfer occurs before

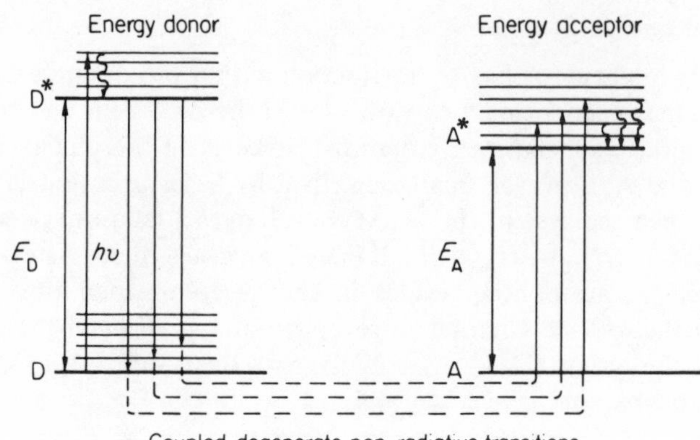

Figure 4.15. Energy level scheme showing the coupling of isoenergetic donor and acceptor transitions necessary for non-radiative energy transfer

emission takes place, and a physical interaction between D* and A is a pre-requisite.

The effect of the coupling between D* and A, represented by the matrix element $\langle\psi_i|H'|\psi_f\rangle$ in equation (4.45), will now be considered. The perturbation H' will contain several terms, among which the most important are electrostatic (Coulombic) and electron exchange interactions, each independently capable of inducing energy transfer.

Coulombic Interaction

This interaction can be expressed as a number of terms, dipole–dipole, dipole–quadripole, multipole–multipole, of decreasing significance. The contribution of the normally dominant dipole–dipole interaction to energy transfer has been treated by Förster, who derives the equation (4.46),

$$k_{ET} = 1{\cdot}25 \times 10^{17}\left(\frac{\phi_E}{n^4\tau_D r^6}\right)\int_0^\infty F_D(\bar{\nu})\varepsilon_A(\bar{\nu})\frac{d\bar{\nu}}{\bar{\nu}^4} \tag{4.46}$$

where ϕ_E is the quantum yield for donor emission, τ_D is the lifetime of the emission, n is the solvent refractive index, and r is the distance in nm between D* and A. $F_D(\bar{\nu})$ is the emission spectrum of the donor, expressed in wavenumbers and normalized to unity (i.e., $\int_0^\infty F_D(\bar{\nu})\,d\bar{\nu} = 1$), and $\varepsilon_A(\bar{\nu})$ is the decadic molar extinction coefficient of A at the wavenumber $\bar{\nu}$.

The distance between D* and A at which energy transfer to A and internal deactivation of D* are equally probable is known as the *critical transfer distance* R_0. Substituting $k_{ET} = \tau_D^{-1}$ into equation (4.46) gives:

$$R_0^6 = 1{\cdot}25 \times 10^{17}\frac{\phi_E}{n^4}\int_0^\infty F_D(\bar{\nu})\varepsilon_A(\bar{\nu})\frac{d\bar{\nu}}{\bar{\nu}^4} \tag{4.47}$$

Introduction of reasonable numerical values into these equations leads to the expectation that k_{ET} can be much larger than k_{diff} and that energy transfer by dipole–dipole interaction can be significant even at distances of the order of 100 Å (10 nm). These expectations have been amply confirmed, as Table 4.1 shows. It is also found, as expected, that k_{ET} is independent of solvent viscosity, except at very low concentrations (*c.* 10^{-4} M) when molecules must diffuse in order to come within the critical transfer distance.

Table 4.1. Long-range Coulombic energy transfer

Donor	Acceptor	R_0 (Å) Theor.	R_0 (Å) Expt.	$10^{-10} k_{ET}$ ($l\,mol^{-1}\,s^{-1}$) Theor.	$10^{-10} k_{ET}$ ($l\,mol^{-1}\,s^{-1}$) Expt.
(a) Anthracene (S_1)	Perylene	31	54	2·3	12
(a) Perylene (S_1)	Rubrene	38	65	2·8	13
(a) 9,10-Dichloro-anthracene (S_1)	Perylene	40	67	1·7	8·0
(a) Anthracene (S_1)	Rubrene	23	39	0·77	3·7
(b) Phenanthrene-d_{10} (T_1)	Rhodamine B	45	47		
(b) Phenanthrene-d_{10} (T_1)	Phenanthrene-d_{10} (T_1)	40	35		
(c) *p*-Phenylbenzalde-hyde (T_1)	Chrysoidin	32	33		
(c) Triphenylamine (T_1)	Chrysoidin	34	52		
(c) Triphenylamine (T_1)	Fuchsin	29	37		

Solvents: (a) benzene at room temperature; (b) cellulose acetate at 77 K; (c) ethanol or dibutyl ether at 77 K. In benzene at room temperature, $k_{diff} \sim 10^{10}\,l\,mol^{-1}\,s^{-1}$. Data taken from Wilkinson *et al.*, ref 24(a), p. 252, and ref. 26.

The only transfer processes allowed by the Coulombic interaction are those in which there is no change in spin in either component. Thus

$$^1D^* + {}^1A \rightarrow {}^1D + {}^1A^*$$

and

$$^1D^* + {}^3A \rightarrow {}^1D + {}^3A^*(T_n)$$

are fully allowed. Triplet–singlet transfer

$$^3D^* + {}^1A \rightarrow {}^1D + {}^1A^*$$

is forbidden. Nevertheless it is observed (see Table 4.1). This is because the spin-forbidden nature of the $^3D^* \rightarrow {}^1D$ transition, whilst strongly reducing k_{ET}, also so prolongs the lifetime of $^3D^*$, particularly in a rigid medium, that the probability of energy transfer can still be high compared with the probability of deactivation of $^3D^*$. For similar reasons, triplet–triplet annihilation has been observed[27] to occur over long distances ($\sim$40 Å) in cellulose acetate films. Long-range energy transfer involving a change in the multiplicity of the acceptor is, of course, not expected to occur.

This discussion has so far centred on the dipole–dipole term in the Coulombic contribution to H'. Dexter[25] has analysed the effect of multipole–multipole interactions, and in particular has shown that for dipole–quadrupole interactions

$$k_{ET} \propto r^{-8}$$

Since this rate constant falls off with distance much more rapidly than that for dipole–dipole transfer, it is likely to be important only at distances $\ll 40$ Å and in those cases where the long-range dipole–dipole transfer is inhibited because $\varepsilon_A(\bar{\nu})$ is small. Note that for Coulombic interactions, k_{ET} is independent of the strength of the optical transition involving D, because the $F_D(\bar{\nu})$ term in equation (4.46) is normalized to unity.

Electron Exchange Interaction

The perturbation H' responsible for energy transfer can include electron exchange terms in addition to the Coulombic terms treated above. The effect of such terms was also analysed by Dexter,[25] who derived the relation (4.48),

$$k_{ET} \propto e^{-2R/L} \int_0^\infty F_D(\bar{\nu})\varepsilon_A(\bar{\nu})\, d\bar{\nu} \tag{4.48}$$

where R is the distance between D* and A, and L is a constant. $F_D(\bar{\nu})$ and $\varepsilon_A(\bar{\nu})$ are the emission and absorption spectra of D and A respectively *both normalized to unity*, so that k_{ET} is independent of the oscillator strength of both transitions (contrast Coulombic interaction).

The negative exponential term in equation (4.48) shows that energy transfer by this mechanism is a short-range phenomenon, which is to be expected since the process involves the interchange of electrons between D* and A and therefore overlap of the orbitals of the two components.

The energy transfer can be formulated thus:

$$D^* + A \longrightarrow (D\cdots A)^* \longrightarrow D + A^* \tag{4.49}$$

Because of the intervention of a bimolecular intermediate, which could be an exciplex or a collision complex, the energy transfer will be subject to conservation of electron spin. Under Wigner's spin rules (p. 101), both of the processes (4.50) and (4.51) are allowed collisionally.

$$^3D^* + {}^1A \longrightarrow {}^1D + {}^3A^* \tag{4.50}$$

$$^1D^* + {}^1A \longrightarrow {}^1D + {}^1A^* \tag{4.51}$$

The former (triplet–triplet energy transfer), though allowed under the exchange interaction, is doubly forbidden under the Coulombic dipole–dipole interaction, while the latter (singlet–singlet energy transfer) is allowed under both interactions. These processes will now be considered in more detail.

Singlet–Singlet Collisional Energy Transfer

The enormous values of the rate constants for long-range dipole–dipole energy transfer and the fact that the rate constant increases with the inverse sixth power of the distance between the donor and acceptor molecules means that *collisional* singlet–singlet energy transfer is likely to be rare and observable only under special conditions. Many of the examples of this phenomenon involve the use of biacetyl as quencher, because the feeble absorption of light by biacetyl over the near ultraviolet and visible regions means that the rate constant for long-range energy transfer will be small (equations 4.44 and 4.46). Exploiting this fact, Dubois and co-workers[28] have demonstrated that the addition of biacetyl to aerated solutions of several aromatic hydrocarbons (benzene, alkyl benzenes and naphthalenes) leads to a quenching of the fluorescence of the hydrocarbon and a sensitization of the fluorescence of biacetyl. The value of k_{ET} obtained from Stern–Volmer analysis was close to k_{diff} calculated from equation (4.22), as expected for exothermic energy transfer.

Triplet–Triplet Energy Transfer (*Triplet Transfer*)

The existence of this phenomenon was first established by Terenin and Ermolaev,[29] who showed that excitation of benzophenone in a rigid glass at 77 K caused phosphorescence of the substrate which, with increasing concentration of added naphthalene, was progressively quenched and replaced by the phosphorescence emission of naphthalene. The light used selectively excited the benzophenone. Since the energy of S_1-naphthalene is greater than that of S_1-benzophenone, energy transfer could not have occurred at the singlet level. Therefore it must have taken place between the triplet states of the two molecules. In other words, the process shown in equation (4.52) must have occurred.

$$Ph_2CO(S_1) \xrightarrow{isc} Ph_2CO(T_1) + C_{10}H_8(S_0) \longrightarrow Ph_2CO(S_0) + C_{10}H_8(T_1) \quad (4.52)$$

Subsequent work[30] has revealed many other examples of the phenomenon in rigid glasses and has established that the energy transfer occurs over a distance of 10–15 Å, comparable with collisional diameters. In 'rigid' glasses it seems that diffusion is rate-determining, for a study[31] of triplet–triplet transfer from phenanthrene-d_{10} to naphthalene-d_8 in hydrocarbon glasses at 77 K shows that $(k_{ET}\eta)$ is a constant over a large range of viscosities, as expected from equations (4.21) and (4.22).

Since biacetyl, exceptionally, phosphoresces strongly in solution, it constitutes a valuable probe for examining triplet–triplet energy transfer in the *fluid* phase. Bäckström and Sandros[32] demonstrated that triplet–triplet transfer occurs from biacetyl as donor to many polynuclear hydrocarbons as acceptors, and they obtained values for both k_{ET} and k_q. Note that $k_q = k_{ET}$ only if the energy transfer is unidirectional. For example, if in the process shown in equation (4.53) the energies of M* and Q* are comparable, then Q*, once formed, can transfer its excitation back to M.

$$M^* + Q \underset{}{\overset{k_{ET}}{\rightleftarrows}} M + Q^* \qquad (4.53)$$

The observed value of k_q will therefore be less than k_{ET}. These points are illustrated in Figure 4.16. Notice that when the energy transfer is more exothermic than about 15 kJ mol^{-1} (3–4 kcal mol^{-1}), $k_q \sim k_{ET} \sim k_{diff}$, and that when energy transfer is endothermic, requiring thermal activation, then k_q and

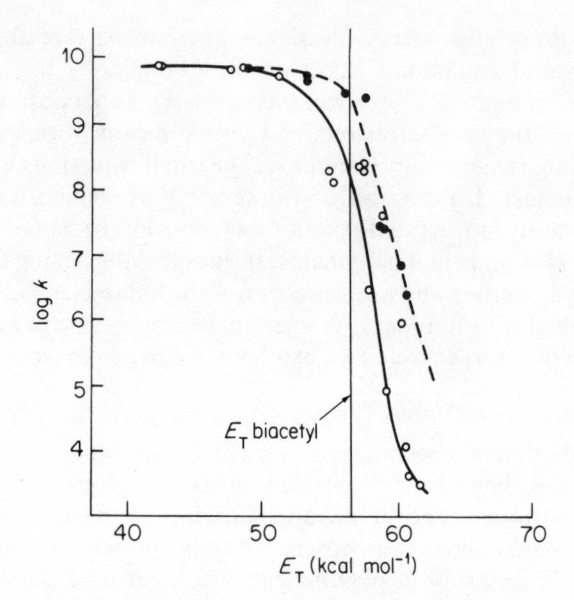

Figure 4.16. Triplet–triplet energy transfer from biacetyl to various acceptors in benzene solutions at room temperature. Rate parameter against triplet energy E_T of acceptor: ○, k_q, observed quenching rate parameter; ●, k_{ET}, energy transfer rate parameter corrected for back transfer. (From A. A. Lamola, *Photochem. Photobiol.*, **8**, 601 (1968), reproduced by permission of Pergamon Press Ltd.)

k_{ET} decrease rapidly, k_q being always less than k_{ET} because of back-transfer. When the triplet energies of M* and Q* are the same, then only the 0—0 bands of the emission spectrum of M* and the absorption spectrum of Q overlap. As predicted by equation (4.48), this leads to a decrease in k_{ET}.

Studies by Porter and Wilkinson[33] on a variety of aromatic hydrocarbons as triplet donors and acceptors, using flash photolysis to monitor donor decay,

Table 4.2. Rate constants for triplet quenching (^3k_q) in hexane solution at room temperature

Donor	Acceptor	Triplet energy gap (kcal mol^{-1}) $E_T(D) - E_T(A)$	^3k_q (l mol^{-1} s^{-1})
Phenanthrene	Iodine	27·7	$1{\cdot}4 \times 10^{10}$
Triphenylene	Naphthalene	6·3	$1{\cdot}3 \times 10^{9}$
Anthracene	Iodine	5·4	$2{\cdot}4 \times 10^{9}$
Phenanthrene	1-Iodonaphthalene	3·1	$7{\cdot}0 \times 10^{9}$
Phenanthrene	1-Bromonaphthalene	2·6	$1{\cdot}5 \times 10^{8}$
Phenanthrene	Naphthalene	0·9	$2{\cdot}9 \times 10^{6}$
Naphthalene	Phenanthrene	−0·9	$\leqslant 2 \times 10^{4}$

established a similar dependence of k_q upon the energy difference between the triplet energies of the donor and acceptor. Some of their data are shown in Table 4.2.

Steric Effects in Collisional Energy Transfer

If the close approach of the partners is essential to collisional energy transfer, then steric effects should be significant, and there are many indications that they are. For example, it has been shown[34] that in the collisional quenching of the fluorescence of the diazabicyclooctene (4.54) in solution by the pairs of dienes (4.55), (4.56) and (4.57), the introduction of *gem*-dimethyl groups reduces k_q by a factor of 3–4.

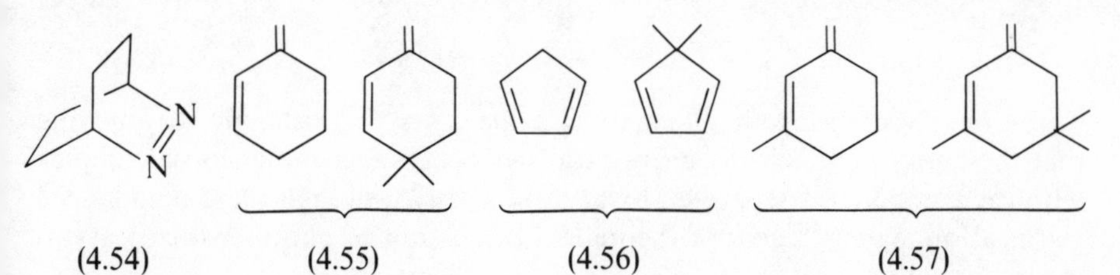

(4.54) (4.55) (4.56) (4.57)

Similarly,[35] in the quenching of benzene fluorescence in the gas phase by simple ketones, increasing α-methyl substitution leads to decreasing values of k_q, as shown in (4.58).

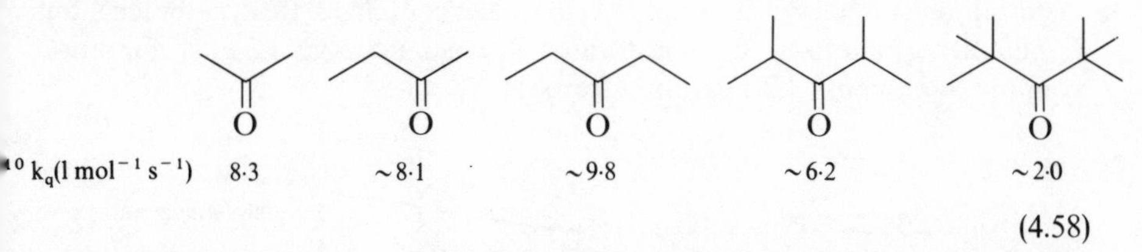

(4.58)

It is also expected that a chiral sensitizer colliding with a quencher molecule would impose its chirality upon the quencher, leading, in suitable cases, to asymmetric induction and optical activity in the products. This was first demonstrated by Hammond and Cole,[36] who used the optically active amide (4.59) to photosensitize the isomerization of *trans*-1,2-diphenylcyclopropane to the *cis*-isomer (4.60). The recovered *trans*-isomer had a specific rotation of +28°, indicating that energy transfer was more efficient to the (−)-enantiomer of the *trans*-isomer than to the (+)-enantiomer.

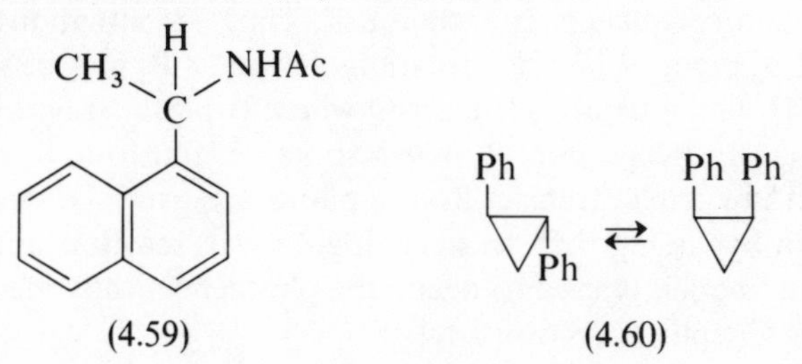

(4.59) (4.60)

Similarly, it has been shown[37] that the optically active steroid (4.61) induces a partial photoresolution of penta-2,3-diene (4.62), and that the amide (4.59) when excited will induce optical activity[38] in the sulphoxide (4.63).

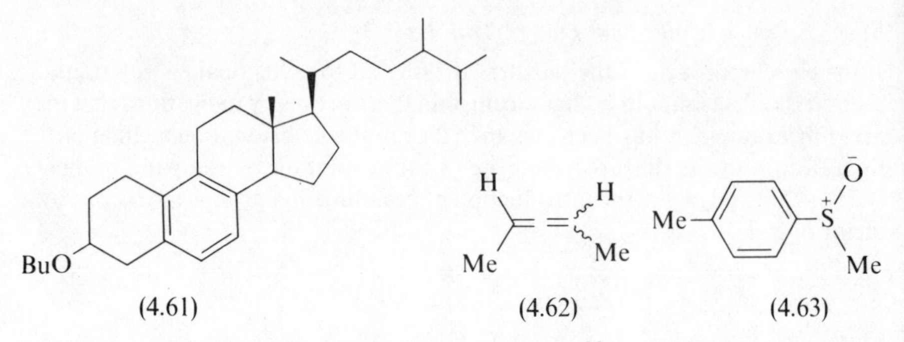

The observed effects are all small, in some cases[39] vanishingly so, implying that the entity in which the energy transfer occurs is a rather loose complex with the components separated by at least a few Å, though these points need clarification. Salem[40] gives a theoretical discussion of photosensitized asymmetric induction.

Photosensitized Processes

It often happens that the S_1 and T_1 states of a molecule have different photochemical reactions. For example, S_1-butadiene cyclizes to cyclobutene but T_1-butadiene dimerizes; β,γ-unsaturated ketones undergo 1,3-acyl migration from the S_1 state but 1,2-migration from the T_1 state.

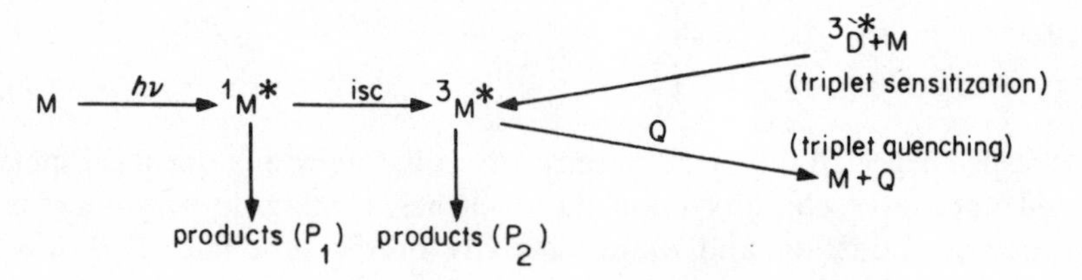

Figure 4.17. Selection between the products P_1 and P_2 may be obtained by quenching or sensitizing $^3M^*$

Control over the nature of the products can be obtained for such systems by sensitization and/or quenching techniques. Thus for situations represented by the scheme of Figure 4.17, the formation of product P_2 may be suppressed by quenching $^3M^*$ with a triplet quencher Q whose triplet energy is less than that of $^3M^*$. Conversely, the product P_2 may be obtained free from P_1 by specifically producing $^3M^*$ by triplet transfer from a photosensitizer $^3D^*$ of higher triplet energy, thereby bypassing $^1M^*$. In an analogous way, sensitization and quenching studies are frequently used to define the photochemically reactive states of molecules (see Chapter 5, section 5.1.4).

For an effective triplet sensitization experiment it is essential that the sensitizer absorb all, or nearly all, of the light and that the triplet energy of the sensitizer be greater than that of the substance being sensitized. Such conditions are satisfied when the energy levels are disposed as in Figure 4.18. The filter removes the short-wavelength radiation required to excite the naphthalene $S_0 \rightarrow S_1$ transition and ensures preferential excitation of the benzophenone, which can transfer its triplet excitation to naphthalene.

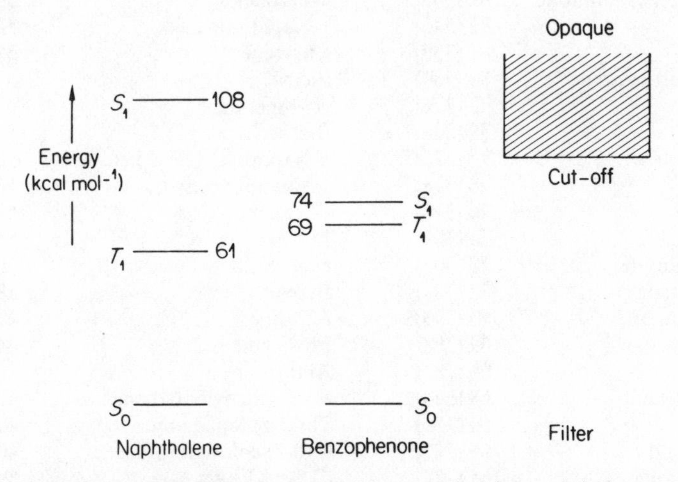

Figure 4.18. Benzophenone as a triplet sensitizer for naphthalene

The desirability of 'bracketing' the S_1 and T_1 levels of the sensitizer by those of the substrate means that ketones, with their small singlet–triplet splitting, high triplet energies and high ϕ_T, make excellent sensitizers.

If the energy levels of the singlet states cannot be disposed as in Figure 4.18, then ultraviolet absorption spectra of the substrate and proposed sensitizer should be examined in order to find a region of the spectrum where the sensitizer absorbs more strongly than the substrate; the irradiation should be conducted at this wavelength and the sensitizer should be used in a sufficiently high concentration to trap most of the available light.

A compilation of triplet energies is given in Table 4.3. Other extensive compilations may be found in references 41 and 42.

Intramolecular Energy Transfer

Intramolecular energy transfer has been observed in many molecules so constructed that the linked D and A moieties cannot collide with each other. Thus Keller[43] has demonstrated complete intramolecular transfer of singlet excitation energy from the naphthalene to the anthrone systems of (4.64) and (4.65), and also transfer of triplet excitation from the anthrone to the naphthalene moiety.

Table 4.3. Triplet energies of sensitizers and quenchers

Compound	E_T kcal(kJ) mol^{-1}	Compound	E_T kcal(kJ) mol^{-1}
Mercury	113 (472)	Isoquinoline	61 (255)
Pyridine	85 (355)	2-Naphthaldehyde	60 (251)
Benzene	84 (351)	2-Naphthyl methyl ketone	59 (247)
Phenol	82 (343)	*trans*-Piperylene	59 (247)
Phenyl methyl sulphone	82 (343)	*p*-Terphenyl	59 (247)
Toluene	82 (343)	1-Naphthoic acid	58 (242)
Anisole	81 (339)	Chrysene	57 (238)
1,4-Dichlorobenzene	80 (334)	Picene	57 (238)
Acetone	78 (326)	*cis*-Piperylene	57 (238)
Aniline	77 (322)	Biacetyl	56 (234)
Benzonitrile	77 (322)	1-Naphthyl methyl ketone	56 (234)
Pyrazine	76 (318)	1-Naphthaldehyde	56 (234)
Benzoin	76 (318)	Coronene	55 (230)
Xanthone	74 (309)	Fluorenone	53 (222)
Diphenylamine	72 (301)	*trans*-Stilbene	50 (209)
Benzaldehyde	72 (301)	Pyrene	48 (201)
Triphenylamine	70 (293)	Acridine	45 (188)
Benzophenone	69 (288)	Phenazine	44 (184)
Fluorene	68 (284)	Anthracene	43 (180)
Triphenylene	68 (284)	1,4-Diphenylbutadiene	42 (176)
Diphenyl	66 (276)	Thiobenzophenone	40 (167)
m-Terphenyl	65 (272)	9,10-Dichloroanthracene	40 (167)
Thioxanthone	65 (272)	Crystal Violet	39 (163)
Diphenylacetylene	63 (263)	Perylene	36 (150)
Anthraquinone	62 (259)	Naphthacene	29 (121)
Quinoline	62 (259)	Oxygen[a]	23 (96)
Phenanthrene	62 (259)		
Michler's Ketone	61 (255)		
Naphthalene	61 (255)		

[a] Energy of $T_1 \rightarrow S_1$ transition.

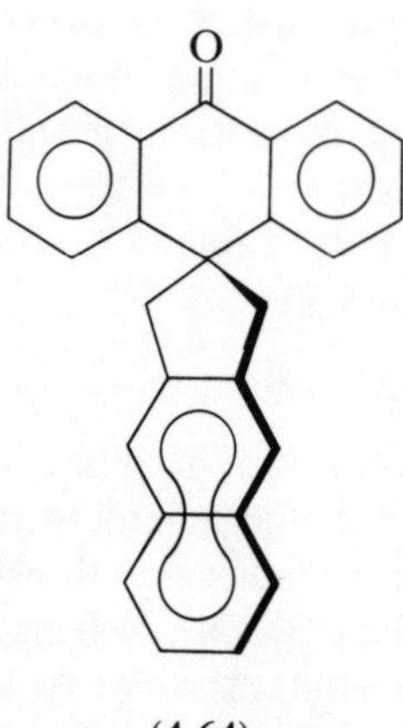

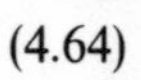
(4.64)

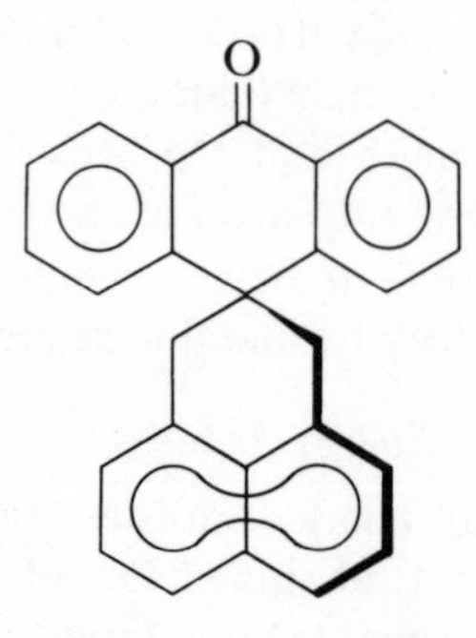

(4.65)

Particularly interesting in this context is the work of Filipescu,[44] who showed that whereas efficient *inter*molecular singlet–singlet energy transfer occurred from (4.66) to (4.67), in (4.68) no *intra*molecular energy transfer, either singlet–singlet or triplet–triplet, could be detected. In (4.68) the two chromophores are

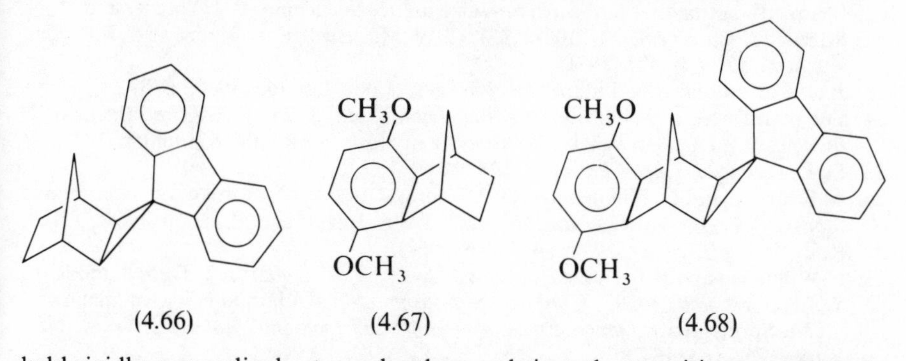

(4.66) (4.67) (4.68)

held rigidly perpendicular to each other, and since the transition moments are perpendicular to each other and to the line joining the centres of the chromophores, the Förster treatment predicts that Coulombic energy transfer will be forbidden. The absence of triplet–triplet transfer from the dimethoxybenzene system to the lower-lying triplet of the fluorene, though the two groupings are only 7 Å apart, suggests that there is an orientational factor associated with triplet–triplet transfer as well. In (4.64) and (4.65) the chromophores are not held *rigidly* perpendicular to each other.

REFERENCES

1. B. Stevens, *Adv. Photochem.*, **8**, 161 (1971); M. Gordon and W. R. Ware (ed.), *The Exciplex*, Academic Press, New York (1975).
2. Th. Förster and K. Kasper, *Z. Phys. Chem. N.F.*, **1**, 275 (1954); *Z. Electrochem.*, **59**, 976 (1955).
3. J. B. Birks, D. J. Dyson and I. H. Munro, *Proc. Roy. Soc.*, **A275**, 575 (1963).
4. J. B. Birks, M. D. Lumb and I. H. Munro, *Proc. Roy. Soc.*, **A280**, 289 (1964); E. Döller and Th. Förster, *Z. Phys. Chem. N.F.*, **34**, 132 (1962).
5. B. Stevens, *Spectrochim. Acta*, **18**, 439 (1962).
6. M. T. Vala, J. Haebig and J. A. Rice, *J. Chem. Phys.*, **43**, 886 (1965).
7. E. Konijnenberg, *Diss.*, Freie Univeriteit Amsterdam, 1963; J. N. Murrell and J. Tanaka, *Mol. Phys.*, **7**, 363 (1964).
8. A. Weller, *Pure Appl. Chem.*, **16**, 115 (1968).
9. H. Beens, H. Knibbe and A. Weller, *J. Chem. Phys.*, **47**, 1183 (1967).
10. See ref. 8, p. 117.
11. H. Knibbe, D. Rehm and A. Weller, *Z. Phys. Chem. N.F.*, **56**, 95 (1967).
12. For references, see ref. 1, p. 207.
13. J. A. Barltrop and H. A. J. Carless, *J. Amer. Chem. Soc.*, **94**, 1951 (1972).
14. W. R. Ware, *J. Phys. Chem.*, **66**, 455 (1962).
15. A. D. Osborne and G. Porter, *Proc. Roy. Soc.*, **A284**, 9 (1965).
16. J. A. Barltrop and R. Owers, *Chem. Commun.*, 1462 (1970).
17. D. Rehm and A. Weller, *Ber. Bunsenges.*, **73**, 834 (1969).
18. B. S. Solomon, C. Steel and A. Weller, *Chem. Commun.*, 927 (1969).
19. T. R. Evans, *J. Amer. Chem. Soc.*, **93**, 2081 (1971).

20. T. Medinger and F. Wilkinson, *Trans. Faraday Soc.*, **61**, 620 (1965); A. R. Horrocks, A. Kearwell, K. Tickle and F. Wilkinson, *ibid.*, **62**, 3393 (1966); A. R. Horrocks and F. Wilkinson, *Proc. Roy. Soc.*, **A306**, 257 (1968).
21. For a review and references see J. B. Birks, *Photophysics of Aromatic Molecules*, Wiley–Interscience, London (1970), p. 496 *et. seq.*
22. It is worth noting that rare-earth ions also induce quenching: P. J. Wagner and H. N. Schott, *J. Phys. Chem.*, **72**, 3702 (1968); G. W. Mushrush, F. L. Minn and N. Filipescu, *J. Chem. Soc.* (*B*), 427 (1971).
23. J. G. Calvert and J. N. Pitts, *Photochemistry*, Wiley, London (1966), p. 89.
24. For reviews see F. Wilkinson, (a) *Adv. Photochem.*, **3**, 241 (1964), and (b) in E. J. Bowen (ed.), *Luminescence in Chemistry*, Van Nostrand Reinhold (1968), p. 154.
25. D. L. Dexter, *J. Chem. Phys.*, **21**, 838 (1953).
26. A. Kearwell and F. Wilkinson in G. M. Burnett and A. M. North (ed.), *Transfer and Storage of Energy by Molecules*, volume 1, Wiley–Interscience, London (1969), p. 129.
27. R. E. Kellogg, *J. Chem. Phys.*, **41**, 3046 (1964).
28. F. Wilkinson and J. T. Dubois, *J. Chem. Phys.*, **39**, 377 (1963); J. T. Dubois and R. L. Van Hemert, *ibid.*, **40**, 923 (1964); B. Stevens and J. T. Dubois in H. P. Kallmann and G. M. Spruch (ed.), *Luminescence of Organic and Inorganic Molecules*, Wiley, New York (1962), p. 115.
29. A. Terenin and V. L. Ermolaev, *Trans. Faraday Soc.*, **52**, 1042 (1956).
30. V. L. Ermolaev, *Opt. Spectrosc.*, **6**, 417 (1959); *Soviet Phys. Uspekhi*, **6**, 333 (1963).
31. B. Smaller, E. C. Avery and J. R. Remko, *J. Chem. Phys.*, **43**, 922 (1965).
32. H. L. J. Bäckström and K. Sandros, *Acta Chem. Scand.*, **12**, 823 (1958); **14**, 48 (1960); K. Sandros and H. L. J. Bäckström, *ibid.*, **16**, 958 (1962); **18**, 2355 (1964).
33. G. Porter and F. Wilkinson, *Proc. Roy. Soc.*, **A264**, 1 (1961); F. Wilkinson, *Adv. Photochem.*, **3**, 241 (1964).
34. T. R. Wright, *D.Phil. Thesis*, Oxford (1970).
35. K. Janda and F. S. Wettack, *J. Amer. Chem. Soc.*, **94**, 305 (1972).
36. G. S. Hammond and R. S. Cole, *J. Amer. Chem. Soc.*, **87**, 3256 (1965).
37. C. S. Drucker, V. G. Toscano and R. G. Weiss, *J. Amer. Chem. Soc.*, **95**, 6482 (1973).
38. G. Balavoine, S. Juge and H. B. Kagan, *Tetrahedron Letters*, 4159 (1973).
39. P. J. Wagner, J. M. McGrath and R. G. Zepp, *J. Amer. Chem. Soc.*, **94**, 6883 (1972).
40. L. Salem, *J. Amer. Chem. Soc.*, **95**, 94 (1973).
41. Ref. 21, p. 256; S. L. Murov, *Handbook of Photochemistry*, Dekker, New York (1973).
42. S. P. McGlynn, T. Azumi and M. Kinoshita, *Molecular Spectroscopy of the Triplet State*, Prentice-Hall, New Jersey (1969), p. 67; D. O. Cowan and R. L. Drisko, *Elements of Organic Photochemistry*, Plenum Press, New York (1976), p. 224.
43. R. A. Keller, *J. Amer. Chem. Soc.*, **90**, 1940 (1968).
44. N. Filipescu in E. C. Lim (ed.), *Molecular Luminescence*, Benjamin, New York (1969), p.697.

Chapter 5

Investigation of Reaction Mechanisms

A great deal of effort has been directed towards the elucidation of the mechanisms of organic photochemical reactions in recent years, and general methods of tackling the problems involved have been developed. Definition of a mechanism for a particular reaction requires identification of excited state intermediates and other short- or long-lived species on the reaction path, and determination of rate constants for the various chemical and physical steps in the sequence. There is obviously a considerable overlap between methods used for investigating photochemical reactions and those used for investigating thermal reactions, especially fast thermal reactions. In this chapter the methods specific to photochemistry, such as the use of excited state quenchers, are dealt with in some detail, and methods of investigation which are common to all mechanistic studies are mentioned and illustrated as appropriate. The material is divided into qualitative methods for identifying reaction products, intermediates and excited states, and quantitative methods for determination of quantum efficiencies and rate constants, but there is no such neat division in practice, and cross-reference is necessary.

5.1 QUALITATIVE METHODS

5.1.1 Products

Identification of the chemical structure of products is the first step in the definition of a reaction, and this is achieved by standard analytical, spectroscopic, chemical and degradative techniques. It is often possible to learn a considerable amount about a mechanistic pathway if the starting material can be 'labelled' in a particular position, and the position of the label in the product can be determined. Isotopic labelling[1] is a valuable method for such investigations since it can reasonably be assumed that the mechanism is substantially the same for reactions of compounds which differ only in the nuclear mass of a small number of atoms. For example, 17-ketosteroids on irradiation in aqueous solution give a reasonable yield of ring-opened carboxylic acid (5.1). The mechanism may involve a direct 4-centre addition of water or may involve a keten as intermediate (5.2), and the use of a deuterium labelled reactant distinguishes between the two pathways.

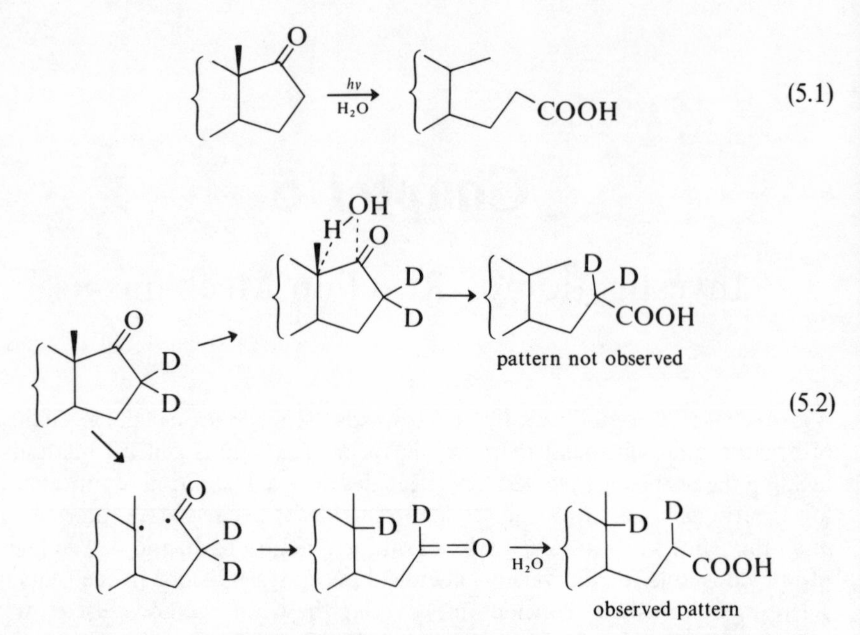

Alternatively, a position in the reactant can be 'labelled' by substitution with an alkyl or other chemically inert group. If no alkyl shift occurs during the course of the reaction, the position of the substituent in the product can be used to differentiate between alternative mechanistic routes, as for the isomerization of the bicylic ketone (5.3).[2]

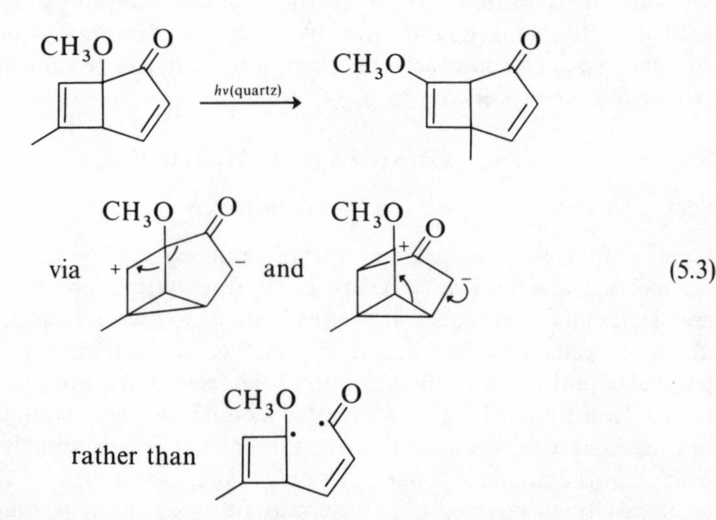

Stereochemical features in the product can often help in a complete definition of a reaction mechanism. If geometrical or optical stereochemistry in the reactant is fully preserved in the product, the mechanism will probably have some concerted character, whereas loss of such stereochemistry in a reaction sequence points to the existence of intermediates which are planar or which exhibit free rotation about single bonds or rapid inversion at asymmetric centres. In principle it is possible that there are two concerted pathways leading at similar rates to isomeric products, and some theoretical support for such a situation in certain cycloaddition reactions has been put forward, but it is likely to be an infrequent occurrence.

5.1.2 Intermediates

One of the advantages of the use of photochemical reactions in synthesis is that the large amount of energy supplied in a specific (electronic) manner can effect conversions to high energy systems which are not readily accessible by thermal processes. These high energy products may be stable under the conditions of reaction, but some are thermally unstable or chemically reactive, and the detection and identification of such intermediates is important in the elucidation of the overall reaction pathway leading to the isolated products. The methods employed depend to a large extent on the lifetime of the species under investigation. For relatively long-lived species ($\tau > 10^1$–10^2 s) it is possible, for instance, to irradiate a sample directly in the cell of a spectrometer. Bands characteristic of the unstable intermediate may then be observed, as for the enol form of acetone (5.4) which is detected when pentan-2-one is irradiated in the cell of an infrared spectrometer.[3]

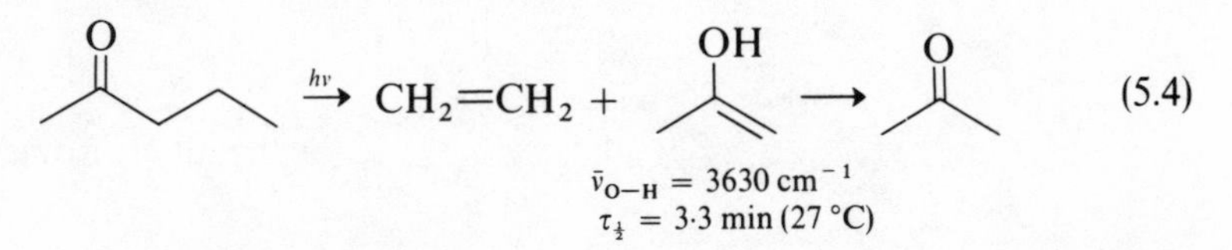

Similarly, in the photoreduction of benzophenone by isopropanol a coloured compound is formed (5.5) whose ultraviolet absorption spectrum can be recorded if the photoreaction is carried out in the cell of an ultraviolet spectrometer, although this may not be an intermediate in the photoreduction.[4]

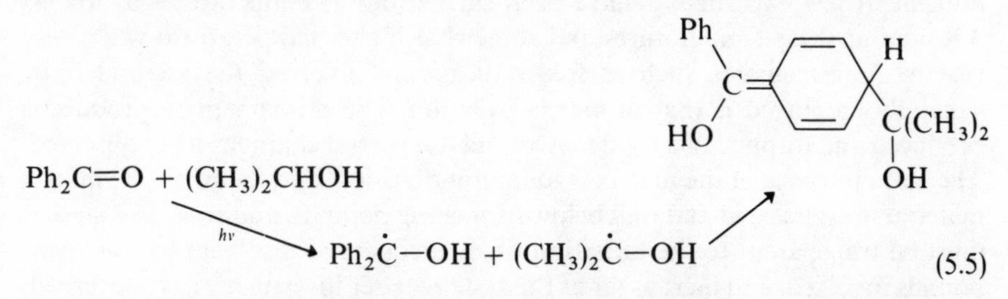

These methods are useful if the lifetime of the intermediate is long enough under normal conditions either for a sufficiently high stationary state concentration to be built up and detected under continuous irradiation conditions, or for a sufficiently high concentration to be achieved for detection immediately after interruption of the irradiation. Extension of the useful range of such methods can be achieved by the use of multiple reflections to give a longer effective pathlength in a spectrometer cell, or by the use of modulated photolysis ('chopping') to increase the signal-to-noise ratio,[5] but the detection of shorter-lived intermediates demands the employment of different techniques.

Low Temperature Photolysis

For intermediates which are short-lived at normal temperatures, it may be possible to conduct a photolysis at low temperature and to determine 'at leisure' the spectral or chemical characteristics of the intermediate formed. This method may lead to characterization of a primary photoproduct and establishment of the subsequent occurrence of a secondary, thermal reaction to give products which are those normally observed on irradiation at room temperature. An example of this is seen in the photochemical isomerization of *trans*- to *cis*-9,10-dihydronaphthalene (5.6). Low-temperature irradiation gives rise to a (10)-annulene which cyclizes to *cis*-9,10-dihydronaphthalene on warming, and it is therefore likely that the room temperature isomerization proceeds by way of the intermediate annulene.[6]

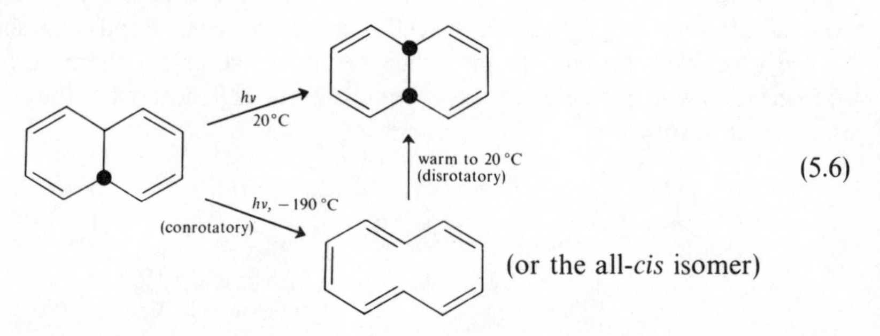

(5.6)

The usefulness of this technique is limited by the temperature dependence of the rate constants for the secondary reactions of the intermediate (i.e., can the thermal reaction be slowed down sufficiently?), and by the temperature limit below which experimental manipulation and observation become too difficult. A few experiments have been carried out at temperatures as low as 4 K, and at these temperatures and somewhat higher temperatures when very reactive intermediates such as free radicals are involved the method most generally employed is that of matrix isolation. The primary photoproduct is produced and trapped in a rigid matrix and its spectral characteristics observed. The main purpose of the matrix is to inhibit diffusion, and therefore the matrix material must be rigid and well below its melting point. In addition the material must be transparent to the radiation used, adequate as a solvent for the compounds involved, and inert as far as the system under investigation is concerned. Nitrogen or argon at 20 K is a good matrix material[7]—xenon often enhances intersystem crossing processes by a heavy atom effect (see Chapter 3, p. 82), and this interference with the photochemical reaction pathway is undesirable in mechanistic studies. Techniques of this type have allowed the identification of cyclobutadiene as a product in the photochemical reaction of the bicyclic lactone from α-pyrone (5.7),[8] and of benzyne in the reaction of the bicyclic dione (5.8).[9]

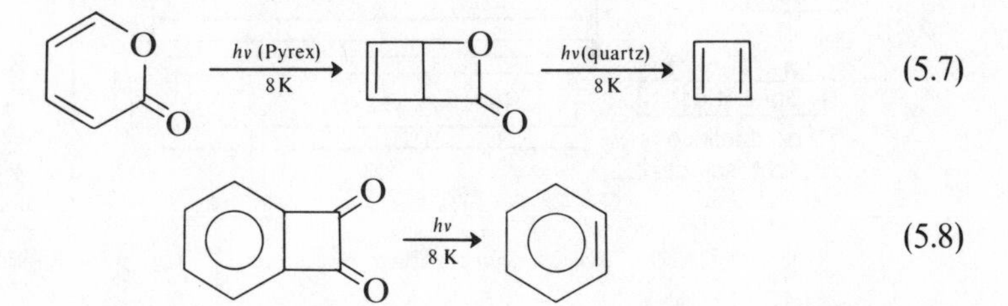

The results of experiments in solid matrices need to be treated with some caution, in the first instance because a matrix environment can affect the spectra of the species produced, and more generally because it is not always possible to extrapolate results obtained for low temperature, solid–state photochemistry to provide information about the same systems in the liquid or vapour phase at ambient temperature. These considerations apart, much valuable information has been obtained from low–temperature studies, and at extremely low temperatures it has been possible to reduce the rate of chemical decay of electronically excited states so that their lifetime is long enough to allow a second photon to be absorbed by a significant proportion of the excited states. The reactions of very high energy excited states thus produced by the absorption of two photons can be studied.

Flash Photolysis

A different approach to the characterization of short-lived or very short-lived intermediates (with lifetimes down to 10^{-11} or 10^{-12} s) is that involved in the applications of flash photolysis.[10] Flash techniques have developed rapidly over the past few years with the introduction of laser sources, and they now represent one of the most powerful tools for studying qualitatively and quantitatively the excited states and intermediates in photochemical processes. In principle flash photolysis involves the generation of a high concentration of a short-lived intermediate using a very high intensity pulse of radiation of very short duration. At a short time interval after the generating pulse the system is analysed, usually by observing its emission or absorption characteristics. A summary of flash techniques is given here, and for greater detail the reader is referred to reviews.[11,12]

A basic flash system in diagrammatic form is shown in Figure 5.1. The conventional flash sources are based on gas discharge lamps (flash durations down to 1 ns), spark discharge sources (flashes down to a few μs) or exploding wire sources (flashes down to a few hundred μs). Detection techniques vary according to the nature of the system and the information required, but in all cases the time resolution of the method is limited by the duration of the initial flash. The emission spectrum of an intermediate can be photographed using a spectrograph, or the visible absorption spectrum can be similarly recorded if an

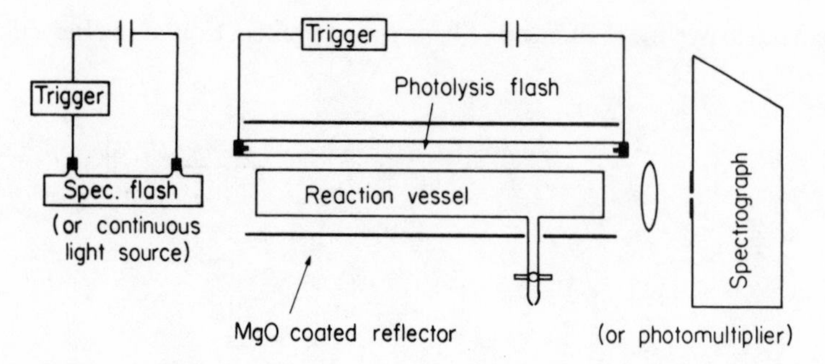

Figure 5.1. Basic flash photolysis arrangement (from D. N. Hague, *Fast Reactions*, (1971), John Wiley and Sons Ltd.)

analytical beam passing through the reaction cell is triggered to flash at a predetermined time interval after the initial flash. Alternatively, the processes can be followed kinetically by monitoring the emission or absorption at a particular wavelength and coupling to an oscilloscope with a time-based sweep. These methods give some indication from the spectrum obtained as to the nature of the species generated, and they allow direct estimation of the lifetime of an intermediate. The difficulties of interpretation lie largely in the possibility of having more than one emitting or absorbing species. In gas phase systems further uncertainties arise because the generating flash can cause a considerable local temperature rise. Short-lived free radicals such as those produced in the photoreduction of aromatic ketones (5.9) are among the many intermediates which can be detected by conventional flash methods.

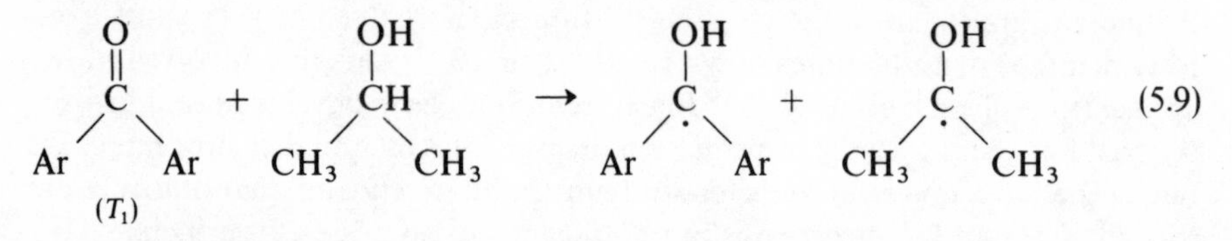

The polychromatic nature of the radiation from conventional gas discharge lamps is a disadvantage which has been largely eliminated by the introduction of lasers for flash photolysis. The radiation from laser sources is unique in its monochromatic nature, and Q-switching allows for the pulse to be of very short duration and highly reproducible (see Chapter 2, section 2.4). A wide range of lasers is available,[13,14] though the most commonly used are solid-state lasers employing rods of such material as ruby or neodymium glass. Often it is necessary to subject the pulse to the process of 'frequency doubling' so that it emerges with a wavelength in the ultraviolet rather than the red or infrared region.

The use of such short duration generating pulses requires a modification of the monitoring procedure for lifetime measurement if any advantage is to be gained over the use of discharge lamps except for monochromaticity. Various methods are employed. An oscilloscope trace of the decay of an intermediate can be obtained if a high intensity (flash discharge) monitoring beam is used to overcome the problems caused by background noise with lower intensity sources. More tediously, background noise can be reduced by repeating the measurement many times and averaging the results, and this can be advantageous if it is important to avoid supplying excess energy to the system under study. It is also possible to build up a picture of the decay of an intermediate by recording a series of spectra at different, predetermined time intervals after the excitation flash or to build up a complete spectrum of the intermediate point by point by changing the setting of the monitoring monochromator in a series of readings at a fixed time interval after the initial flash. Laser flash photolysis has been particularly widely used to study the excited singlet states of aromatic hydrocarbons in solution.[12]

A further stage in the improvement of time resolution in laser studies is achieved by the use of mode-locked laser sources, which provide a chain of laser pulses each of a few picoseconds (10^{-12} s) duration. Techniques using these sources enable processes with rate constants around 10^{11}–10^{12} s^{-1} to be studied, and this limit is sufficiently high for vibrational relaxation processes to be measured.

CIDNP

A recently developed technique for studying products formed in rapid radical combination steps is based on the phenomenon of chemically induced dynamic nuclear spin polarization (CIDNP).[15] A spin-paired product from combination of two radicals is formed initially in a non-equilibrium distribution of nuclear spin states, and the intensity of microwave absorption in an applied magnetic field is different from that normally observed for the compound. The effect is investigated by irradiating a sample in the cavity of an n.m.r. spectrometer, and the intensity of the signals for the protons attached to the atoms forming the new bond in the product is observed. CIDNP effects are also observed with other nuclei such as carbon-13. The signals may be of higher or lower intensity than normal, and may be negative (i.e., microwave emission occurs). The 'normal' spectrum is obtained within a few seconds of interrupting the illumination. As an example, on irradiation of benzophenone in ethylbenzene, part of the n.m.r. signal for the —CH— proton in the alcohol produced (5.10) is negative under conditions of steady illumination (see Figure 5.2).[16]

$$Ph_2C{=}O + PhCH_2CH_3 \xrightarrow{h\nu} Ph_2\dot{C}{-}OH + Ph\dot{C}H{-}CH_3 \rightarrow Ph_2C(OH){-}CH(CH_3){-}Ph \qquad (5.10)$$

The best description of the origin of the effect is as a consequence of electron spin polarization accompanying interaction of radical pairs within a solvent cage.[17] Singlet–triplet mixing occurs via hyperfine interaction in radical pairs, and this is coupled to the nuclear spin states. Singlet–triplet transitions, which

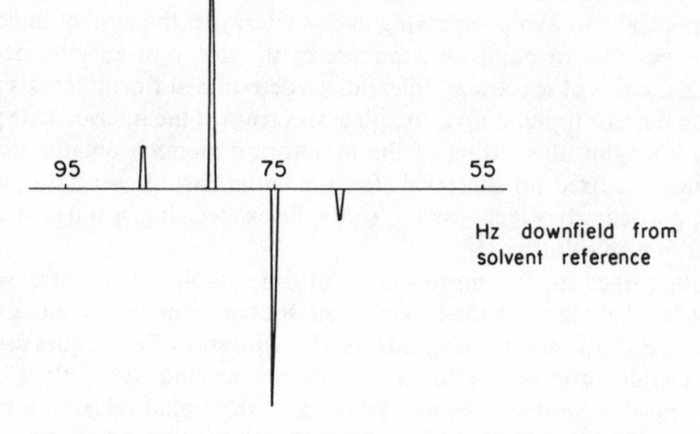

Figure 5.2. Adapted from G. L. Closs and L. E. Closs, *J. Amer. Chem. Soc.*, **91**, 4550 (1969), and reproduced by permission of the American Chemical Society. CIDNP spectrum from the irradiation of benzophenone in ethylbenzene. Copyright by the American Chemical Society

are involved because of the need for a singlet radical pair for combination, therefore occur with nuclear-spin dependent probability. Subsequent radical reaction rates depend on the extent of singlet–triplet mixing, and this results in selective population of nuclear spin configurations in the product.

It is possible to distinguish between products formed within the solvent cage and those formed subsequent to radical separation, and also between the spin multiplicities of different precursors to the radical pairs if the signs of the nuclear spin coupling constants are known. The spectrum observed differs according to whether the radical pair is formed from a singlet state, from a triplet state, or from a coming together of radicals produced separately. Such a difference is observed in the spectra of 1,1,2-triphenylethane produced by the three methods shown (5.11) for generating phenylmethyl and diphenylmethyl radicals.[18] Rules for the qualitative interpretation of spectra in terms of the state of the radical pair reactants (singlet, triplet or free, and cage or non-cage), electron–electron exchange coupling constant, and electron-nuclear hyperfine splitting have been developed.[19]

$$\begin{aligned} \mathrm{Ph_2CH{-}N{=}N{-}CH_2Ph} &\xrightarrow{h\nu} \mathrm{Ph_2CH^{\cdot} + PhCH_2^{\cdot}} \quad \text{(singlet)} \\ \mathrm{Ph_2C{=}\overset{+}{N}{=}\overset{-}{N} + PhCH_3} &\xrightarrow{\text{heat}} \mathrm{Ph_2CH^{\cdot} + PhCH_2^{\cdot}} \quad \text{(triplet)} \\ \mathrm{Ph_2CH_2 + PhCH_3} &\xrightarrow{\text{peroxide}} \mathrm{Ph_2CH^{\cdot} + PhCH_2^{\cdot}} \quad \text{(separate)} \end{aligned} \tag{5.11}$$

The technique has been used to study a number of photochemical reactions which occur by way of free radicals, including radical ions produced by electron transfer in such reactions as those between amines and the excited states of ketones (5.12).[20] Biradicals can also be investigated, although for effective singlet–triplet mixing the radical centres must be at least five atoms apart. The major drawback of CIDNP is its great sensitivity, so that there is the danger that an observed effect may represent only a minor part of the product-forming reaction pathway, and that some signals may be caused by a minor by-product.

$$Ar_2C{=}O + R_3N \xrightarrow{h\nu} Ar_2C{=}O^{\cdot -} + R_3N^{\cdot +} \qquad (5.12)$$

More recently, electron spin resonance emission has been observed[21] from photochemically produced radicals, and this CIDEP (chemically induced dynamic electron spin polarization) can provide information about the triplet state and its relaxation. The effect is thought to arise from initial unequal intersystem crossing rates to the sub-levels of the triplet excited state.

Trapping

The existence of short-lived intermediates can often be inferred from the results of experiments using chemical 'trapping' agents. These are selected to be reactive towards the suspected intermediate but unreactive towards the starting material, products, or any other intermediate or excited state in the reaction sequence. The basic principle is the same as for excited state quenching (see Chapter 5, p. 144) in that inhibition of the formation of normal products and the appearance of other products derived from the trapping agent can be taken as evidence for a particular intermediate, about which information can be gathered by characterization of the 'new' products.

The intermediacy of short-lived free radicals in a photochemical process can be demonstrated in favourable cases by converting the radical to a long-lived nitroxide radical using an alkyl nitroso-compound (5.13)[22] or nitrone as radical scavenger, or by conversion to a fully electron-paired species using iodine, an alkanethiol (5.14) or a stable free radical such as nitric oxide (5.15) or diphenylpicrylhydrazyl. The long-lived products can be identified at leisure, and this identification in conjunction with the observed inhibition of the photochemical reaction provides strong evidence for a reaction pathway in a particular system.

$$R^{\cdot} + Bu^{t}{-}N{=}O \rightarrow (R)(Bu^{t})N{-}O^{\cdot} \qquad (5.13)$$

$$R^{\cdot} + Bu{-}SH \rightarrow R{-}H + Bu{-}S^{\cdot} \qquad (5.14)$$

$$R^{\cdot} + NO^{\cdot} \rightarrow R{-}NO \qquad (5.15)$$

An attempt to trap radical intermediates can complicate a situation if the trapping agent or the product of trapping is photochemically active, and there is always the possibility that the scavenger reacts with an electronically excited state of the reagent rather than with a subsequent intermediate to produce the observed quenching of reaction. This seems to be a more commonly encountered situation than has sometimes been imagined. For instance, nitroxide radicals such as Bu^t_2NO can trap radical intermediates and also quench excited states. A cautious approach to the interpretation of results from radical trapping experiments is required.

Different types of chemical trapping agent can be employed according to the chemical nature of the suspected intermediate. For instance, *o*-methylbenzophenone is not photoreduced efficiently by isopropanol, whereas the *m*- and *p*-isomers and *o*-*tert*-butylbenzophenone undergo efficient reduction under similar conditions. The unreactivity of the *o*-isomer is attributed to the rapid and reversible formation of a photo-enol in an intramolecular hydrogen abstraction step (5.16). The enol can be trapped with added dienophile such as maleic anhydride, and the structure of the adduct is consistent with the proposed reaction scheme.[23]

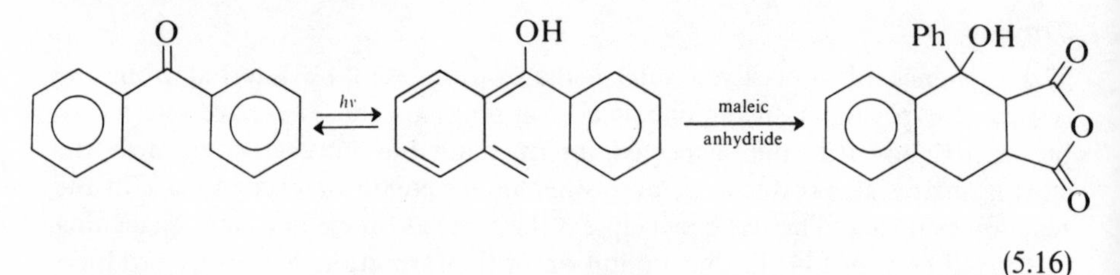

(5.16)

5.1.3 Lowest Energy Excited States

The excited state responsible for photochemical reaction is often, though not always, either the lowest excited singlet state or the lowest triplet state of the reactant, and characterization of these states is based on spectroscopic data.

The longest wavelength band normally seen in the ultraviolet absorption spectrum arises from singlet ⟶ singlet absorption and corresponds to the lowest energy electronic transition. The nature of this transition is usually apparent from the magnitude of the extinction coefficient, the effect of solvent polarity on the position of the maximum (more strictly, on the position of the 0—0 vibrational band, since the band shape may vary with solvent), and the direction of polarization of the absorption band (see Chapter 2, p. 24). For $n \rightarrow \pi^*$ transitions ε_{max} is usually small (10^1–10^2 l mol^{-1} cm^{-1}), the band shows a large 'blue shift' (i.e., a shift to shorter wavelength) in a more polar solvent, and the transition is polarized in a direction perpendicular to the molecular plane. For fully allowed $\pi \rightarrow \pi^*$ transitions in alkenes, dienes or carbonyl compounds ε_{max} is large (10^3–10^4 l mol^{-1} cm^{-1}), the band shows a small 'red shift' (i.e., a shift to longer wavelength) in a more polar solvent, and the direction of polarization is parallel to the molecular plane. For intramolecular charge-transfer transitions, which can be regarded as extreme cases of $\pi \rightarrow \pi^*$ transitions, ε_{max} is very large (10^4–10^5 l mol^{-1} cm^{-1}), and the band usually shows a pronounced red shift as the solvent polarity increases.

It is not normally possible to observe singlet–triplet absorption spectra by conventional techniques because of the very low extinction coefficient associated with these transitions. The intensity of $S_0 \rightarrow T_1(\pi, \pi^*)$ absorption can be increased by the use of high pressures of added oxygen, and triplet (π, π^*) energies have been determined for conjugated dienes using this method.[24] Triplets other than (π, π^*) which may be of lower energy are not detected. Another technique which enhances the relative intensity of $S \rightarrow T$ transitions is that of phosphorescence excitation (see Chapter 3, p. 71), in which the variation of the intensity of phosphorescence emission at a given wavelength is monitored as a function of the wavelength of the exciting radiation. The method has greater sensitivity than normal absorption techniques, and it is usually possible to observe the singlet $\rightarrow$ triplet absorption bands. Aromatic ketones such as acetophenone show two singlet $\rightarrow$ triplet bands, one corresponding to a $n \rightarrow \pi^*$ and one to a $\pi \rightarrow \pi^*$ transition, which can be distinguished by the fact that the latter is enhanced in intensity by a heavy-atom solvent (see Chapter 2, p. 24).

Decay processes of the excited states of a molecule which involve emission of visible or ultraviolet radiation have already been described, as well as the methods of identifying the emitting state (see Chapter 2, p. 61). Emission spectra sometimes show more detailed fine structure than absorption spectra, and it is possible to assign the nature and energy of the emitting state. This is normally the lowest energy excited state of the molecule, either singlet or triplet, although exceptions are known in which emission originates from a higher excited state—azulene, for instance, fluoresces from the second excited singlet state.

5.1.4 Reactive Excited States

The methods described in the previous section are for the identification of the lowest energy or emitting excited states of a molecule, but these may not be the states directly responsible for photochemical reaction. Higher energy or non-emitting states may be involved. Sensitization and quenching methods are the most generally useful for characterization of reacting states. The theoretical aspects of energy transfer are described in Chapter 4, applications in qualitative mechanistic studies are dealt with in this section, and applications in quantitative investigations of reaction mechanism in the next section.

Most qualitative studies have involved triplet sensitizers or triplet quenchers, partly because the results of such studies are more easily interpreted than those involving singlet sensitizers and quenchers (singlet state sensitizers and quenchers are usually also triplet state sensitizers and quenchers, but the reverse is not true). Triplet sensitizers must be chosen carefully for use with a particular system.[25] The sensitizer must absorb radiation strongly at the wavelength employed so that absorption by the substrate is minimized. The sensitizer must have a high quantum efficiency for intersystem crossing, and its triplet state so formed must be long-lived so that transfer of its energy to another molecule can be efficient. This means that there must be no rapid loss of electronic energy by intersystem crossing to ground state, nor must there be rapid photochemical reaction in the triplet state of the sensitizer. On this latter count dibenzyl ketone $(PhCH_2)_2C{=}O$ is ruled out as a high energy (330 kJ mol^{-1}, 79 kcal mol^{-1}) triplet sensitizer because it undergoes a very rapid bond-cleavage reaction in the triplet state. The most widely used compounds which are good triplet sensitizers without at the same time being singlet sensitizers are aromatic ketones such as acetophenone or benzophenone.

The qualitative use of triplet sensitizers is of value in two respects. First, if triplet sensitization leads to products which are different from those obtained on direct irradiation (i.e., in the absence of sensitizer), then the reactive state in the direct irradiation is not the same as the triplet state obtained by sensitization. If triplet sensitization produces the same products as direct irradiation, then it is proved that reaction *can* occur through the triplet state. If in addition the ratio of different products is the same whether the reaction is sensitized or not, then it is very likely that both sensitized and unsensitized reactions proceed through the same excited state.

Secondly, if a reaction can be triplet sensitized it may be possible to determine the approximate triplet energy of the reactive state by using a range of sensitizers of different triplet energies. Normally those sensitizers with an energy greater than that of the reactive state lead to relatively efficient reactions, and those with an energy lower than that of the reactive state lead only inefficiently to products. The changeover in efficiency occurs at about a triplet energy equal to that of the reactive state of the substrate. The value so obtained is approximate, and in the region where sensitizer and substrate have similar energies reverse energy transfer from substrate triplet state to sensitizer ground state can occur efficiently, and this is manifest in a dependence of the efficiency of sensitization on the concentration of sensitizer. Self-quenching by the sensitizer shows itself in the same way, and this process is particularly important for ketones with (π, π^*) triplet states, such as thioxanthone, when used as sensitizers. In assigning triplet energies by this method it should be remembered that sensitizers may be able to act through a higher excited state than the lowest or emitting state. Anthracene is one such compound, which acts as a triplet sensitizer through an upper excited triplet state.[26]

Quenching techniques can in a similar manner yield information about the multiplicity and energy of the reactive excited state in a photochemical reaction. Sensitization and quenching are different aspects of the same phenomenon, namely energy transfer. The process is termed sensitization when it is described from the point of view of the energy acceptor and its reactions, and is termed quenching when described from the point of view of the energy donor and its reactions.

The most generally useful triplet state quenchers are conjugated dienes, which have relatively low triplet state energies (220–250 kJ mol^{-1}, 53–59 kcal mol^{-1}) and lowest singlet states of much higher energy (>400 kJ mol^{-1}, 100 kcal mol^{-1}) so that singlet quenching is of less importance. If the photochemical reaction of a compound can be quenched efficiently by a known triplet quencher, then there is strong evidence for a reactive triplet excited state. Again, it may be possible to obtain an approximate value for the energy of the reactive triplet state by using a range of quenchers of different triplet energy and noting the region of triplet energy in which the efficiency of quenching changes markedly.

Singlet quenching and mixed singlet/triplet quenching techniques can be employed in mechanistic studies, and a use of the latter is seen[27] in the study of the photochemistry of hexan-2-one (5.17). Only part of the photochemical formation of acetone can be quenched by 1,3-diene triplet quenchers, but in the presence of excess 1,3-diene the remainder of the reaction can be quenched by added biacetyl which is able to quench both triplet and singlet states of the ketone. It is apparent, therefore, that both singlet and triplet states of hexan-2-one play a part in the production of acetone.

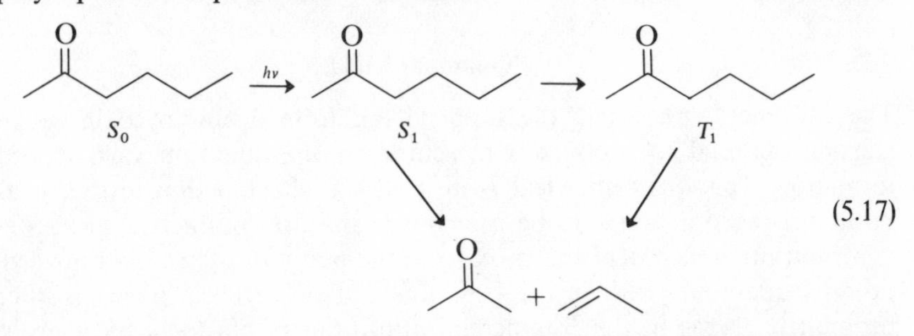

(5.17)

A comparison of the effect of quencher on product yield with the effect on the intensity of luminescence (fluorescence or phosphorescence) can be particularly useful in identifying a reactive excited state. If a quencher reduces equally the quantum yield of product formation and the quantum yield of luminescence under the same conditions, then it is likely that the emitting state is a precursor of the photochemical product. If the quencher has no effect on light emission, the emitting state is not the immediate precursor for the chemical reaction. Similarly, if the quenching of a photochemical reaction is accompanied by fluorescence or phosphorescence emission from the quencher, the quenched state can be identified as a precursor (though not necessarily the immediate precursor) of the product.

Flash photolysis techniques, described in an earlier section, provide a powerful method for detecting excited state intermediates in a photochemical reaction and for determining their lifetimes. The methods can be adapted to measure a lifetime in the presence of excited state quencher, and this gives a direct measure of the rate constant for the quenching process. Such measurements carried out for a range of quenchers enable a more precise assignment of excited state energies to be made, and the understanding of some aspects of the mechanism of the *cis–trans* isomerization of stilbene has been aided considerably in this way.[28]

5.2 QUANTITATIVE METHODS

A more complete description of a reaction mechanism than an identification of the final products and of the intermediates on the reaction pathway requires quantitative data on the efficiency of reaction and on the rate constants for the steps involved, particularly for the primary processes undergone by the electronic excited states of the reactant. The processes are often very fast under normal conditions (first-order rate constants greater than $10^9\ s^{-1}$ are not unusual), and to some extent the general methods of investigating fast thermal reactions can be applied to the study of photochemical processes. In particular, competition methods have been extensively used in which the rate constants for excited state reactions are estimated by comparison with known rate constants for alternative bimolecular processes, usually energy transfer processes. Direct measurement of excited state lifetimes, especially by flash photolytic methods, is becoming increasingly important.

5.2.1 Quantum Yields

The efficiency with which the supplied radiation is employed in converting starting material to product is measured by the quantum yield of product formation. This quantum yield is in itself a useful function in that it allows optimum reaction times to be calculated, and for commercial processes the contribution of electrical energy costs to the unit cost requires a knowledge of product quantum yield. In the U.K. this cost of electrical power is often the prohibitive factor in the commercial utilization of photochemical processes, and at present only processes with very high quantum yield (such as the photo-initiated chlorination of alkanes) or processes for which the energy cost is small by comparison with reagent costs (as in the sequence from corticosterone to aldosterone) are economically viable.

The quantum yield (ϕ) for formation of a particular product is given by:

$$\phi = \frac{\text{amount of product formed (in moles or molecules)}}{\text{amount of radiation absorbed (in einsteins or photons)}}$$

$$= \frac{\text{rate of formation of product}}{\text{rate (i.e., intensity) of absorption of radiation}}$$

The quantum yield can be related to the rate constants in a particular system, but normally it is not a useful direct measure of the reactivity of an excited state, that is, of the rate constant for the primary chemical step from the excited state. This is because the quantum yield is a ratio of rate constants and sums of rate constants. For the simple system (5.18) a steady-state analysis (5.19) enables an expression for the quantum yield for formation of product B to be derived (5.20).

$$A \xrightarrow[\text{rate } I]{h\nu} A^* \begin{cases} \xrightarrow{k_r} B \\ \xrightarrow{k_{-1}} A \end{cases} \tag{5.18}$$

$$\frac{d[A^*]}{dt} = I - (k_r + k_{-1})[A^*] = 0 \tag{5.19}$$

$$\phi_B = \frac{k_r[A^*]}{I} = \frac{k_r}{k_r + k_{-1}} \tag{5.20}$$

The quantum yield measurement provides a value for k_r (a true measure of reactivity) only if the excited state lifetime $[\tau_0 = (k_r + k_{-1})^{-1}]$ is known.

The measurement of quantum yields is, in principle, straightforward, and requires the measurement of the product formed in a given time and the radiation absorbed by the system in that time. The estimation of stable products presents no difficulties beyond those normally encountered in chemical analysis, although for accurate work it is best to restrict the extent of reaction (to less than 5% or even less than 2%), and this may necessitate particular care in analysis. This is one reason for the widespread use of gas chromatography in quantitative photochemical studies. The direct measurement of radiation absorbed presents greater problems. In the first instance the radiation is required to be monochromatic, or to contain only a narrow range of wavelengths, since quantum yields are sometimes wavelength-dependent, and, more importantly, it is unusual for a reagent and an actinometer system to be equally sensitive to the same range of radiation wavelengths. The use of chemical filter solutions, glass filters or interference filters allows the selection of individual bands from a low or medium pressure mercury arc source.[29] Conventional grating or prism monochromators can be used, but they have the disadvantage of having a very low power output at any particular wavelength. Whatever method is chosen for isolating a narrow band of radiation wavelengths, it is possible that the source output will vary with time, even when a stabilized power supply is used, and for high accuracy a beam-splitter (e.g., a partially-silvered mirror) should be used so that photolysis and actinometry can be carried out simultaneously (Figure 5.3).

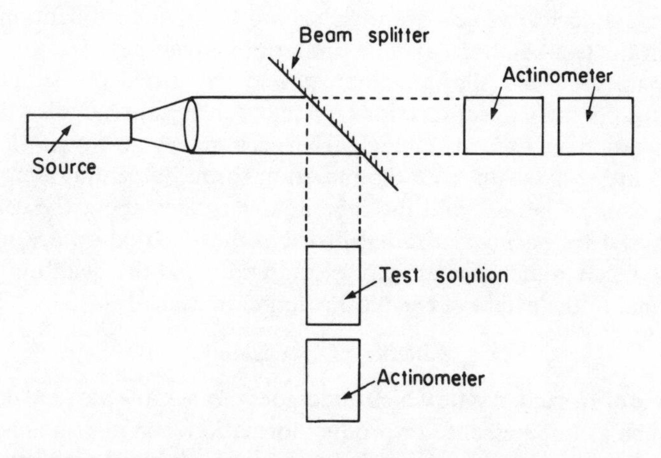

Figure 5.3. A beam splitter used to allow simultaneous photolysis and actinometry

'Direct' physical measurement of radiation intensity[30] can be carried out by radiometric methods (thermopiles, thermistors or bolometers), or by photoelectric methods (phototubes, or photovoltaic cells), but these techniques are not widely used in routine measurement, largely because of constructional difficulties in that the measuring device should ideally be sited in the exact position normally occupied by the photolysis cell (unless a beam-splitter is employed) and it should be such a shape and of such material that the radiation incident on it and recorded by the instrument is the same as that absorbed by the solution under photolysis. These conditions are difficult to achieve satisfactorily, and it is more usual to measure light intensity by chemical actinometry. This employs as a secondary standard a chemical reaction whose quantum yield is known and can be reproduced accurately and which is fairly constant over a useful range of wavelengths (or whose variation with wavelength is known). This type of actinometer is readily adapted to the same photolysis cell as the sample under investigation. Quantum yield measurement then requires only two chemical analyses, one for reaction product and one for actinometer product. The most commonly used liquid phase chemical actinometers employ transition metal oxalates which undergo redox reactions on irradiation, such as potassium trisoxalatoferrate(III), uranyl oxalate, or vanadium(V) iron(III) oxalate.[31] For vapour phase reactions the production of carbon monoxide from acetone is a common standard system.

It is often advantageous to use as chemical actinometer a reaction which has itself been calibrated against a 'primary' chemical actinometer standard. For instance, in a study of the photochemistry of an alicyclic ketone the photoreaction of hexan-2-one to give acetone (estimated by g.l.c.) may be used as actinometer. The value of this is (i) that the actinometer solution has absorption characteristics as near as possible to those of the solution under study, (ii) that the solvent is the same in the two solutions so that errors caused by reflectance at the glass/solvent interface are minimized (these errors can be particularly high when cylindrical vessels are used[32]), and (iii) that the quantum yields of reaction in the two solutions may be similar in magnitude.

An apparatus which allows actinometry to be carried out simultaneously with several photochemical reactions (or photochemical reaction with a range of concentrations of excited state quencher) is a 'merry-go-round' (Figure 5.4). The tubes are equidistant from the radiation source (a central lamp or a surrounding bank of lamps), and the tube holder rotates about the central axis. Errors caused by an uneven radial distribution of radiation are minimized. Each tube receives the same amount of radiation, and the quantum yields are proportional to the extent of chemical reaction in each.[33]

5.2.2 Kinetics of Quenching

The lifetime of an excited state which undergoes measurable physical or chemical change (such as luminescence or product formation) can in principle be determined by the employment of a suitable 'quencher' which interacts with the excited state in an energy transfer process whose rate constant is known. This

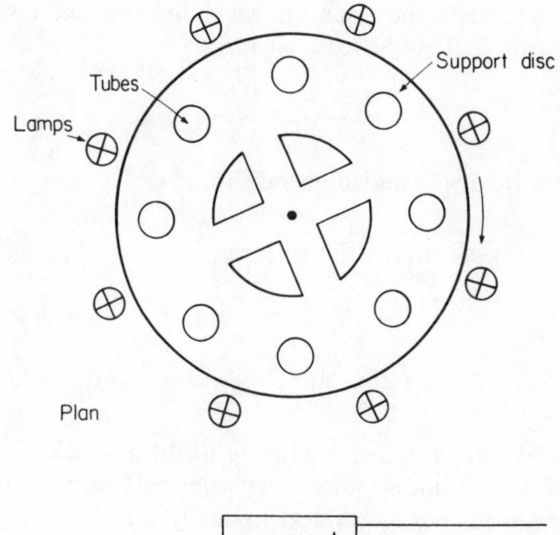

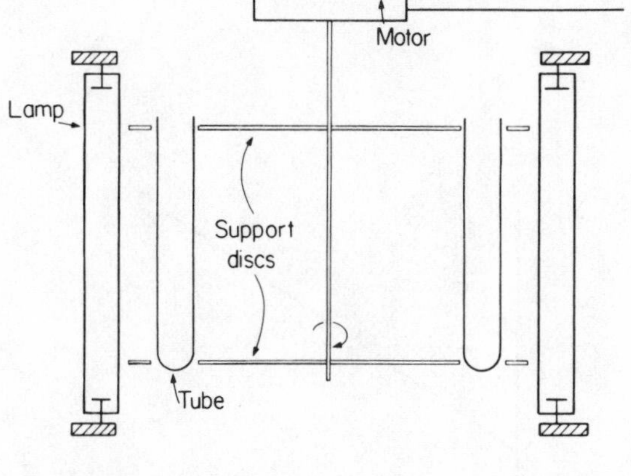

Figure 5.4. 'Merry-go-round' apparatus

is illustrated for the simple system (5.21) in which a single chemical product (B) is formed in a unimolecular reaction from the first-formed excited state (A*) of the starting material, and in which A* can be quenched by an added species (Q).

$$\begin{aligned} A &\xrightarrow{h\nu} A^* && \text{rate } I\ (= \text{intensity}) \\ A^* &\rightarrow A && k_{-1}\,[A^*] \\ A^* &\rightarrow B && k_r\,[A^*] \\ A^* + Q &\rightarrow A + Q^* && k_q\,[A^*][Q] \end{aligned} \qquad (5.21)$$

k_{-1} is a composite first-order rate constant which takes into account radiative and radiationless deactivation of A*. Using the steady-state approximation (cf. 5.19), the quantum yield for product formation in the absence of quencher (ϕ_B^0) can be expressed as:

$$\phi_B^0 = \frac{k_r}{k_r + k_{-1}}$$

and the quantum yield with added quencher (ϕ_B) as:

$$\phi_B = \frac{k_r}{k_r + k_{-1} + k_q[Q]}$$

Hence

$$\frac{\phi_B^0}{\phi_B} = 1 + \frac{k_q[Q]}{k_r + k_{-1}} = 1 + k_q\tau_0[Q] \quad (5.22)$$

where τ_0 is the lifetime of the excited state in the absence of quencher. This expression is linear in [Q] and is the form of the normal 'Stern–Volmer' quenching plot[34] often obtained in photochemical studies (Figure 5.5).

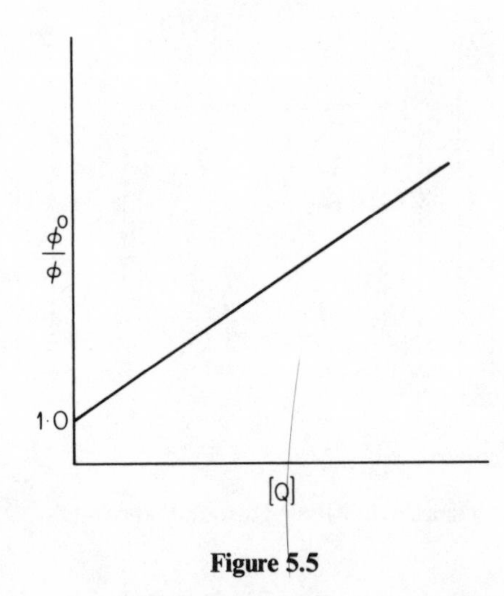

Figure 5.5

The assumptions made in the derivation of (5.22), apart from those always implicit in the application of the steady-state treatment, are (i) that the conditions are such that the intensity (I) of radiation absorbed by the reagent is effectively unchanged over the period of reaction, (ii) that the quencher interacts only with the reactive excited state involved in the product-forming process, it interacts only by energy transfer, and its concentration is not significantly altered by chemical reaction, and (iii) that the product is formed from only one excited state of A, which is the one quenched by Q. These assumptions will be considered in turn.

For the intensity of radiation absorbed by A to be constant over the period of reaction it is necessary either that the reduction in [A] occasioned by reaction is insignificant by comparison with the original concentration of A, or that A has a much higher extinction coefficient at the wavelength employed than any products formed and is present in sufficient concentration to absorb all (99–100%) of the radiation throughout the reaction. The use of low percentage conversion is usually preferred, and this has the advantage of discouraging interference by the reaction products with the processes in the photochemical reaction pathway. Products and quencher do sometimes absorb a significant amount of the incident radiation, and it is possible to apply a correction for situations where quencher absorbs a part of the radiation or where the percentage conversion is such that absorption by products builds up significantly during the course of the reaction. The second method of ensuring a constant high percentage absorption by the reagent is to use a sufficiently high initial concentration of A for absorption due to A to be greater than 99% throughout the reaction. This must not be taken too far, or another source of error becomes apparent: if the absorbance of the solution is very high the bulk of the incident radiation is absorbed in the first fraction of the pathlength of the reaction cell, and product builds up in this volume if diffusion or mixing within the cell is not rapid. This inhomogeneous generation of product can affect quantitative results. The problems associated with absorption of the incident radiation include also the differences which arise because of reflectance at the surfaces of cells of different shape or containing different solvents.[32]

If the quencher interacts with more than one excited state of A, the effect on product formation may be different from that expressed in (5.22). Interaction of quencher with an excited state formed by non-radiative transition *from* the reactive (product-forming) excited state does not affect product formation unless a serious depletion of Q in a chemical reaction ensues. Interaction with an excited state earlier on the reaction pathway than the reactive excited state does have an effect. Such a situation (5.23) leads to the quenching expression given in (5.24).

$$\begin{array}{rcll}
A & \xrightarrow{h\nu} & A^* & \text{rate } I \\
A^* & \rightarrow & A & k'_{-1}[A^*] \\
A^* & \rightarrow & A^{\ddagger} & k_{ic}[A^*] \\
A^* + Q & \rightarrow & A + Q^* & k'_q[A^*][Q] \\
A^{\ddagger} & \rightarrow & A & k_{-1}[A^{\ddagger}] \\
A^{\ddagger} & \rightarrow & B & k_r[A^{\ddagger}] \\
A^{\ddagger} + Q & \rightarrow & A + Q^* & k_q[A^{\ddagger}][Q]
\end{array} \tag{5.23}$$

In the absence of quencher,

$$\phi_B^0 = \frac{k_r}{k_r + k_{-1}} \cdot \frac{k_{ic}}{k_{ic} + k'_{-1}}$$

and with added quencher,

$$\phi_B = \frac{k_r}{k_r + k_{-1} + k_q[Q]} \cdot \frac{k_{ic}}{k_{ic} + k'_{-1} + k'_q[Q]}$$

Hence,

$$\frac{\phi_B^0}{\phi_B} = \left(1 + \frac{k_q[Q]}{k_r + k_{-1}}\right)\left(1 + \frac{k'_q[Q]}{k_{ic} + k'_{-1}}\right)$$
$$= (1 + k_q\tau_0[Q])(1 + k'_q\tau'_0[Q]) \qquad (5.24)$$

This expression is quadratic in [Q], and the form of such a quenching curve is shown in Figure 5.6. This situation is frequently encountered when a reaction occurs from a triplet state and a quencher is employed which quenches both the triplet state and the singlet state from which the triplet state arises by inter-system crossing.

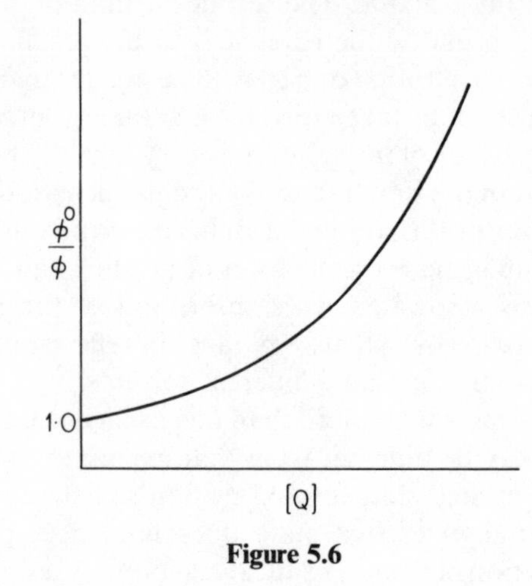

Figure 5.6

To examine the third assumption made in the simple scheme (5.21) it is necessary to consider the situation in which product is formed from two excited states of A, one or both of which may be affected by the quencher. The complete scheme (5.25) gives rise to the general expression (5.26).

$$\begin{array}{rcll}
A & \xrightarrow{h\nu} & A^* & \text{rate } I \\
A^* & \rightarrow & A & k'_{-1}[A^*] \\
A^* & \rightarrow & A^{\ddagger} & k_{ic}[A^*] \\
A^* & \rightarrow & B & k'_r[A^*] \\
A^* + Q & \rightarrow & A + Q^* & k'_q[A^*][Q] \\
A^{\ddagger} & \rightarrow & A & k_{-1}[A^{\ddagger}] \\
A^{\ddagger} & \rightarrow & B & k_r[A^{\ddagger}] \\
A^{\ddagger} + Q & \rightarrow & A + Q^* & k_q[A^{\ddagger}][Q]
\end{array} \qquad (5.25)$$

In the absence of quencher,

$$\phi_B^0 = \frac{1}{(k_{ic} + k'_r + k'_{-1})}\left(k'_r + \frac{k_r k_{ic}}{k_r + k_{-1}}\right)$$

With added quencher,

$$\phi_B = \frac{1}{(k_{ic} + k'_r + k'_{-1} + k'_q[Q])}\left(k'_r + \frac{k_r k_{ic}}{k_r + k_{-1} + k_q[Q]}\right)$$

Hence,

$$\frac{\phi_B^0}{\phi_B} = (1 + k'_q\tau'_0[Q])(1 + k_q\tau_0[Q])\left(\frac{k'_r + k_r k_{ic}\tau_0}{k'_r(1 + k_q\tau_0[Q] + k_r k_{ic}\tau_0)}\right) \quad (5.26)$$

where $\tau'_0 = (k_{ic} + k'_r + k'_{-1})^{-1}$ and $\tau_0 = (k_r + k_{-1})^{-1}$.

If $k'_q = 0$, that is if the quencher affects only the second of the reactive states, the quenching curve tends to a horizontal asymptote given by

$$\frac{\phi_B^0}{\phi_B} = \left(1 + \frac{k_r k_{ic}\tau_0}{k'_r}\right)$$

This can also be expressed as

$$\frac{\phi_B^0}{\phi_B} = \left(1 + \frac{(\varphi_B^0)}{(\varphi_B^0)'}\right)$$

where (φ_B^0) and $(\varphi_B^0)'$ are the contributions to ϕ_B^0 from states $A^{\ddagger}$ and A^* respectively.

The form of such a quenching curve is shown in Figure 5.7. This is a situation commonly encountered when a photochemical reaction occurs through both singlet and triplet excited states of a compound and the system is studied using a triplet state quencher.

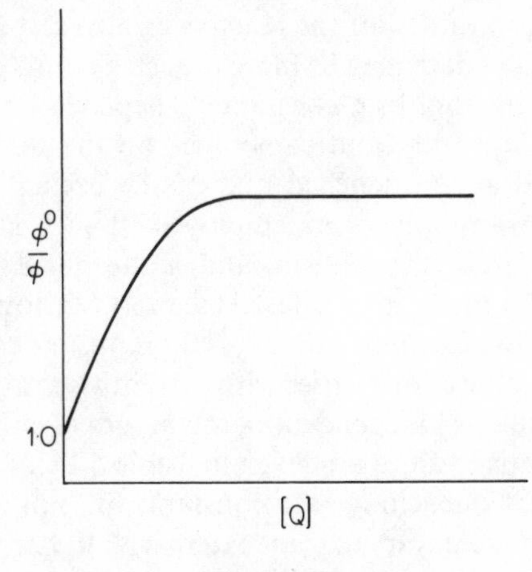

Figure 5.7

If both excited states are involved and both are quenched, the full expression (5.26) applies, and the quenching curve may take different forms. The gradient may decrease from its initial value and approach a constant positive non-zero value (Figure 5.8), or it may increase from its initial value and approach a

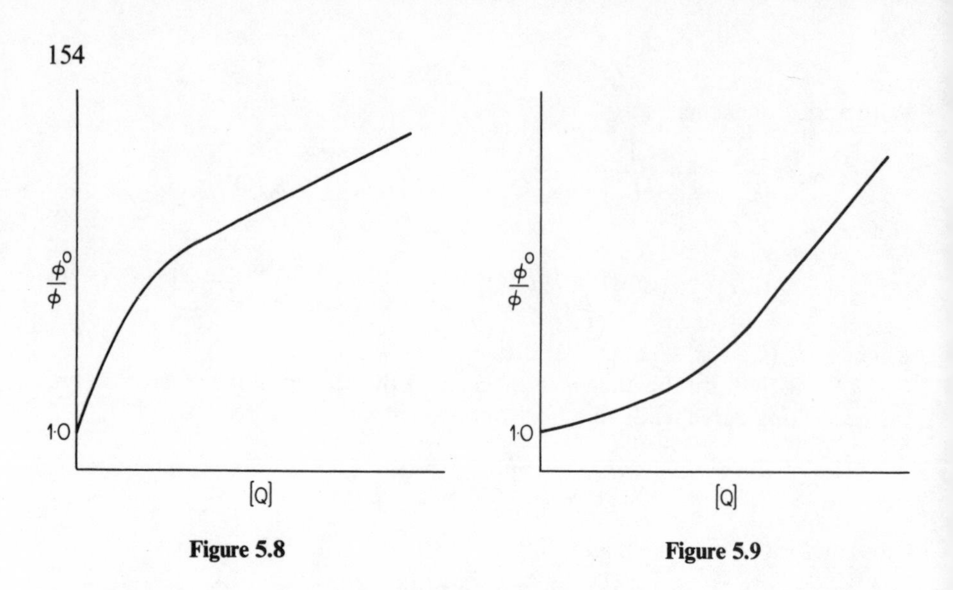

Figure 5.8 **Figure 5.9**

constant positive non-zero value (Figure 5.9) according to whether $k'_q\tau'_0(1 + (\varphi^0_B)/(\varphi^0_B)')$ is less than or greater than $k_q\tau_0$. It is possible for a linear plot to be obtained if $k'_q\tau'_0(1 + (\varphi^0_B)/(\varphi^0_B)') = k_q\tau_0$, and such a situation has already been reported. Its diagnosis is not difficult, because a different quencher can be found which will interact with only one excited state, and this will differentiate between this complex situation and the very simple one where only excited state is involved and quenched.

It should be apparent from the above treatment that the interpretation of the results of quenching data should be approached with care, but that if the data are sufficiently precise, $k_q\tau_0$ values for the reactive excited state(s) can be extracted. A statistical treatment of data may be more accurate than a graphical method.[35] The value of the quenching rate constant k_q depends on the mechanism by which the quencher operates, and sometimes on the difference between the energy of the excited state quenched and of the excited state to which the quencher is raised. Many quenchers employed in such studies operate by a collisional energy transfer mechanism, and if the quencher is such that its excited state energy is more than a few kcal mol^{-1} below that of the donor excited state, then it can be assumed that k_q is the diffusion-controlled quenching rate constant.[36] The value of this varies with solvent viscosity and with temperature, but on the whole it is independent of the nature of either the donor or the quencher. Representative values are given in Table 5.1.

Diffusion-controlled quenching rate constants are not always known with great precision, and they can vary to some extent with the structures of the species involved, and therefore absolute lifetime values calculated on this basis are subject to some error. The relative lifetimes for a series of similar compounds can be obtained with greater confidence because the value of k_q will not vary greatly from one compound to another in a series. Certain quenchers, particularly *cisoid* conjugated dienes such as cyclopentadiene, seem to quench at rates in excess of those expected on the basis of the 'normal' k_q values. This may reflect

Table 5.1. Values for the diffusion-controlled quenching rate constant (k_q)

Solvent	k_q/l mol^{-1} s^{-1}	*a*
Benzene	5×10^9	*b*
Pentane	13×10^9	*b*
Hexane	11×10^9	*b*
Hexadecane	$4{\cdot}5 \times 10^9$	*b*
t-Butanol	2×10^9	*c*
Acetonitrile	11×10^9	*c*

[a] At 25 °C.
[b] P. J. Wagner and I. Kochevar, *J. Amer. Chem. Soc.*, **90**, 2232 (1968).
[c] P. J. Wagner, *J. Amer. Chem. Soc.*, **89**, 5898 (1967).

a molecular geometry which is particularly favourable for collisional energy transfer. Another possible source of error arises at high concentrations of quencher, when k_q is higher than the low-concentration value because of a 'nearest neighbour' quenching effect.[37] This arises because the excited state is often formed with a quencher molecule in its immediate environment if there is a high concentration of quencher in the solution. Diffusion is therefore not required for energy transfer to take place, and the overall rate constant for quenching is higher than the diffusion-controlled value. This effect shows itself as a deviation from linearity at high concentration in a Stern–Volmer quenching plot (see 5.22 and Figure 5.5).

Up to this point it has been assumed that only one quencher is involved, but as described in an earlier section (see Chapter 5, p. 145) 'double' quenching experiments have been usefully employed in which one reactive excited state of A is totally quenched by excess quencher, and the quenching of the second reactive excited state is then studied using a second quencher which is more effective than the first for this particular excited state.

5.2.3 Excited State Lifetimes

The quenching techniques outlined in the previous section provide values of excited state lifetimes τ. Such lifetimes can also be measured directly either by following the luminescence decay in conventional emission spectroscopy (see Chapter 3, section 3.1.4), or by following absorption or emission decay using flash photolysis techniques (see Chapter 5, p. 137). For conventional methods to be useful the intensity of fluorescence or phosphorescence must be high enough to be monitored with reasonable accuracy, and the measurements should be made under the same reaction conditions as for other quantitative data with which the τ value is to be combined. This is particularly relevant to phosphorescence studies, since the most usual conditions for measurement of phosphorescence lifetime involve a rigid glass at liquid nitrogen temperature. A lifetime under these conditions is often much longer than the lifetime measured in liquid solution at room temperature. This difference is attributed to the presence of very small but significant amounts of quenching impurities or to the variation with temperature of rate constants for chemical reactions of the excited state, particularly reaction with the solvent.

The measured lifetime of an excited state can provide an estimate of the rate constant for reaction in the excited state. The lifetime is defined by the expression:

$$\tau = (k_r + k_{-1})^{-1} \tag{5.27}$$

where k_r is the rate constant for chemical reaction and k_{-1} is the sum of rate constants for physical decay processes (phosphorescence and intersystem crossing to ground state for a lowest triplet state, or fluorescence, radiationless decay to ground state and intersystem crossing to triplet state for a lowest excited singlet state). In those instances where an estimate of k_{-1} can be made, a value can be assigned for k_r from equation (5.27). The most general way of estimating rate constants (k_x) for emission and for $S \rightarrow T$ intersystem crossing is to measure the quantum efficiency (ϕ_x) for a process and to combine this value with the measured lifetime (τ) according to equation (5.28).

$$\phi_x = \tau k_x \tag{5.28}$$

Quantum efficiencies for formation of triplet states can be estimated by 'triplet counting' methods in which excess triplet quencher is employed and the number of triplet states formed is calculated from the number of product molecules arising from chemical reaction of the triplet state of the quencher. For example, through its triplet state an acyclic conjugated diene undergoes *cis–trans* isomerization and gives dimers, and for a given diene the quantum efficiencies for these processes can be measured.[38] When excess diene is employed to quench a triplet state in an unknown system, the extent of chemical reaction of the diene provides a direct measure of the extent of formation of the triplet state of the donor, and hence a measure of the quantum efficiency for triplet formation. The isomerization of *cis*-penta-1,3-diene and the dimerization of cyclohexa-1,3-diene are the two most commonly used triplet counters.

The rate constant (k_r) for chemical reaction of an excited state is a direct measure of the reactivity of that state. It is normally obtained indirectly from values of excited state lifetime and other rate constants. Because a chemical reaction is often not a simple one-step process it is therefore not always possible to estimate the rate constant for reaction from the quantum yield of product formation according to the expression (5.29).

$$\phi_B = \tau k_r \tag{5.29}$$

This relationship breaks down if there is more than one pathway open to any intermediate on the reaction pathway from the electronic excited state to the product, and it holds only if the product is formed in a single-step process or in a series of consecutive steps each (after the first) occurring with unit efficiency.

Photochemical reactions very often involve intermediates which can undergo reaction to regenerate starting material. As an example, the intramolecular elimination reaction of the triplet state of aromatic ketones is a two-stage process, and after the initial hydrogen abstraction has taken place, a reverse hydrogen transfer can occur to regenerate the ground state of the ketone (5.30). In this system the value of the rate constant for the primary reaction step is greater than that predicted by equation (5.29) because some of the biradical intermediate has not given rise to the measured product.

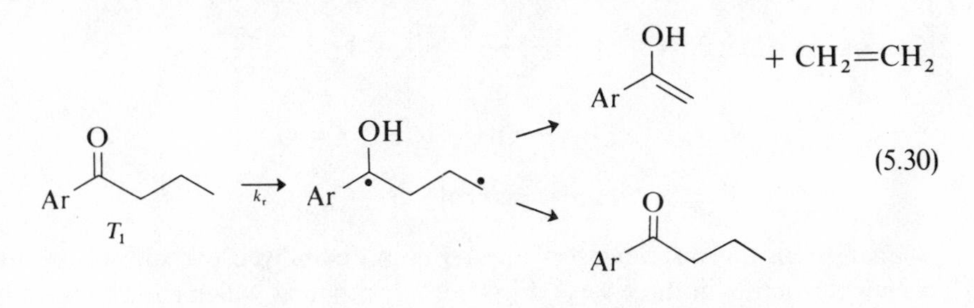

(5.30)

5.2.4 Controlled Variables

The establishment of a photochemical reaction mechanism can be aided by the study of the effects on the reaction of controlled changes in the intensity or nature of the radiation employed or in the pressure or temperature of the system.

The effect of intensity can be used to determine the number of quanta of radiation required to effect a particular chemical change. Under 'normal' circumstances, where one molecule of product arises from only one singly excited state of the reactant, the rate of formation of the product is directly proportional to the intensity of the absorbed radiation. A different dependence on intensity can arise in various ways. First, if the reactive excited state is a 'doubly' excited state which requires two photons for its formation, then the rate of formation of product is proportional to the square of the intensity. Reactions from such excited states have not been widely reported, but they do occur when a high concentration of first excited state can be generated, either by the use of a laser source or by the use of low temperature matrices in which the excited state lifetime is relatively long.

A second situation in which the rate of product formation is proportional to the square of the intensity arises when two singly excited states are involved in the formation of one molecule of product. This is not usually the direct interaction of two excited states, but rather of two intermediates formed from separate excited states. On this basis a distinction can be made between two of the proposed schemes for the formation of benzpinacol by photoreduction of benzophenone in propan-2-ol (5.31). The observed dependence of the rate of product formation on the first power of the intensity rules out the first of the mechanisms.

$$\left.\begin{array}{l}\mathrm{Ph_2C{=}O} \xrightarrow{I} \mathrm{Ph_2C{=}O^*} \xrightarrow{\mathrm{(CH_3)_2CHOH}} \\ \qquad \mathrm{Ph_2\dot{C}{-}OH + (CH_3)_2\dot{C}{-}OH} \\ \mathrm{2Ph_2\dot{C}{-}OH} \rightarrow \text{pinacol}\end{array}\right\} \text{rate} \propto I^2.$$

$$\left.\begin{array}{l}\mathrm{Ph_2C{=}O} \xrightarrow{I} \mathrm{Ph_2C{=}O^*} \xrightarrow{\mathrm{(CH_3)_2CHOH}} \\ \qquad \mathrm{Ph_2\dot{C}{-}OH + (CH_3)_2\dot{C}{-}OH} \\ \mathrm{(CH_3)_2\dot{C}{-}OH + Ph_2C{=}O} \longrightarrow \\ \qquad \mathrm{Ph_2\dot{C}{-}OH + (CH_3)_2C{=}O} \\ \mathrm{2Ph_2\dot{C}{-}OH} \rightarrow \text{pinacol}\end{array}\right\} \text{rate} \propto I \qquad (5.31)$$

Changes in the wavelength of the radiation employed can affect a photochemical reaction in three ways. First, for a compound which has two or more accessible (singlet) excited states the initial population of states depends on the wavelength of the exciting radiation, and in many instances it is possible to populate one excited state selectively. This may have no effect on the photochemistry, because an initially formed higher energy excited state can be rapidly deactivated to give the lowest energy state from which reaction then occurs. However, if a chemical reaction of the higher energy state can compete effectively with deactivation, a wavelength effect will be observed. Several systems are known which exhibit this type of wavelength-dependence of chemical products, such as the cycloaddition reaction of thioketones with certain alkenes (5.32). One type of product is obtained with radiation of wavelength > 500 nm, but a different type of product with 366 nm radiation.

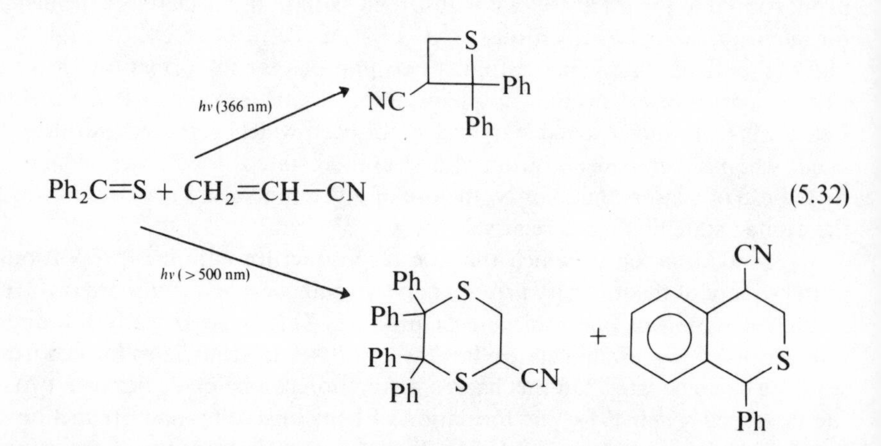

Secondly, a change in wavelength may lead to a change in the observed major products because the first-formed product is itself photochemically reactive. For example,[39] 1,2-dihydronaphthalene gives a tricyclic compound with radiation of wavelength 400 nm, but if 280 nm radiation is employed a conjugated pentaene can be isolated, and this is converted more slowly to the tricyclic product (5.33). At 280 nm the dihydronaphthalene has the higher extinction coefficient, and this enables a significant concentration of pentaene to build up.

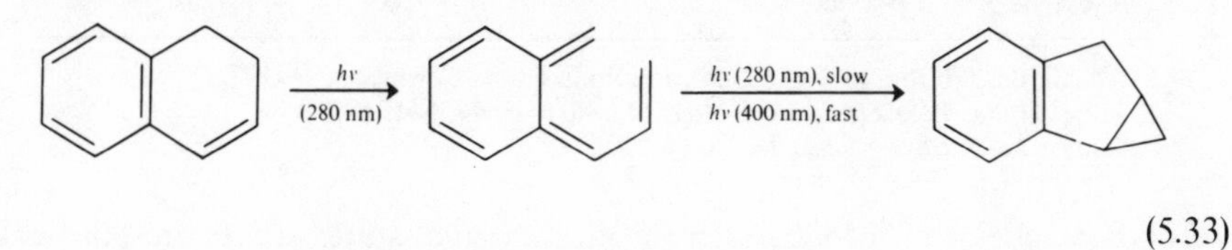

(5.33)

A third effect of wavelength is manifest particularly in gas phase reactions, where collisional deactivation is (relatively) slow. Most chemical processes occur more rapidly from a higher vibrational level than from the lowest vibrational level, and this is true of excited state reactions as of ground state reactions. The effect of decreasing wavelength of excitation in this respect is to populate initially a higher vibrational state of the electronic excited state, and under conditions where chemical reaction does compete effectively with vibrational deactivation this leads to an increase in rate constant and in quantum yield for reaction. This type of behaviour is very common in gas phase reactions where the rates of collision are relatively low, but it is not usual in solution because of the much greater rates of collision with solvent molecules and therefore of vibrational deactivation.

A related effect is that of pressure on gas phase reactions, since increasing the pressure of a reagent (or of increasing total pressure by means of an inert additive) increases the rate of collision and hence the rate and efficiency of vibrational deactivation. Changing pressure therefore causes changes in rate constants and quantum yields for photochemical reactions, and this again is a widely reported effect. As a simple example,[40] the conversion of methyl isonitrile to acetonitrile (5.34) occurs with a quantum efficiency of 1.4 at 10 mmHg pressure, but the quantum yield is too small to be measured at a pressure of 1 atm. A similar large effect can be achieved for the same reason by a change of phase from vapour to solution, and whereas photolysis of acetone gives carbon monoxide with high quantum efficiency in the gas phase, very little CO is produced in solution.

$$CH_3NC \xrightarrow{h\nu} CH_3CN \qquad (5.34)$$

A change in temperature can appreciably affect rate constants and quantum efficiencies for photochemical reactions. It seems that radiative and radiationless decay processes of excited states are generally not greatly affected by changes in temperature (for instance, the phosphorescence lifetime of a photochemically inactive compound is much the same whether measured at room temperature in very pure inert fluid solution or at 77 K in a glass matrix[41]), but primary and

Table 5.2. Arrhenius activation energies for photochemical reactions

Reaction	E_a/kJ mol^{-1}	(kcal mol^{-1})
Ph—CO—$(CH_2)_2CH_3$ → Ph—CO—CH_3	28	(6·6)[a]
Ph—CO—$(CH_2)_3CH_3$ → Ph—CO—CH_3	20	(4·7)[a]
Ph—CH=CH—Ph *trans* → *cis*	15	(3·5)[b]
m-Xylene → *o*/*p*-xylenes	20	(4·7)[c]

[a] J. C. Scaiano, J. Grotewald and C. M. Previtali, *Chem. Commun.*, 391 (1972).
[b] J. Saltiel and J. T. D'Agostino, *J. Amer. Chem. Soc.*, **94**, 6445 (1972).
[c] D. Anderson, *J. Phys. Chem.*, **74**, 1686 (1970).

secondary chemical processes from the excited state are often markedly temperature-dependent. The experimental dependence can be expressed in the form of a normal Arrhenius equation $k = A\,e^{-E/RT}$, and because fast or very fast physical decay processes are available to excited states their observable chemical reactions will also be fast. The activation energies normally encountered for primary processes in photochemistry are therefore low. A few representative values are listed in Table 5.2. For many of the earlier quantitative studies of photochemical reactions in solution the temperature is not specified, and only recently has the importance of temperature control in quantitative work been realized.

REFERENCES

1. J. S. Swenton in E. Buncel and C. C. Lee (ed.), *Isotopes in Organic Chemistry*, Elsevier, Amsterdam (1975), chapter 5.
2. W. G. Dauben, K. Koch, S. L. Smith and O. L. Chapman, *J. Amer. Chem. Soc.*, **85**, 2616 (1963).
3. G. R. McMillan, J. G. Calvert and J. N. Pitts, *J. Amer. Chem. Soc.*, **86**, 3602 (1964); similar enols can be generated in high concentration at low temperature (< -40 °C), A. Henne and H. Fischer, *Angew. Chem. Intern. Ed.*, **15**, 435 (1976).
4. J. Chilton, L. Giering and C. Steel, *J. Amer. Chem. Soc.*, **98**, 1865 (1976).
5. P. B. Ayscough and R. C. Sealy, *J. Photochem.*, **1**, 83 (1972/3).
6. E. E. van Tamelen, T. L. Burkoth and R. H. Greeley, *J. Amer. Chem. Soc.*, **93**, 6120 (1971).
7. E. L. Wehry in J. M. Fitzgerald (ed.), *Analytical Photochemistry and Photochemical Analysis*, Marcel Dekker, New York (1971), p. 184; J. K. Burdett and J. J. Turner in M. Moskovits and G. A. Ozin (ed.), *Cryochemistry*, Wiley, New York (1976), p. 493.
8. O. L. Chapman, C. L. McIntosh and J. Pacansky, *J. Amer. Chem. Soc.*, **95**, 614 (1973).
9. O. L. Chapman, C. L. McIntosh, J. Pacansky, G. V. Calder and G. Orr, *J. Amer. Chem. Soc.*, **95**, 6134 (1973).
10. F. W. Willets in K. R. Jennings and R. B. Cundall (eds.), *Progress in Reaction Kinetics*, Vol. 6, Pergamon, Oxford (1971), p. 51.
11. G. A. Porter and M. A. West in G. G. Hammes (ed.), *Investigation of Rates and Mechanism of Reactions*, Part II, 3rd edition, Wiley–Interscience, New York (1974), chapter 10.
12. M. A. West in W. R. Ware (ed.), *Creation and Detection of the Excited State*, Volume 4, Dekker, New York (1976), p. 217.
13. F. H. Fry in J. M. Fitzgerald (ed.), *Analytical Photochemistry and Photochemical Analysis*, Marcel Dekker, New York (1971), p. 163.

14. C. B. Moore, *Ann. Rev. Phys. Chem.*, **22**, 387 (1971).
15. J. K. S. Wan and A. J. Elliott, *Accounts Chem. Res.*, **10**, 161 (1977); A. R. Lepley and G. L. Closs (ed.), *Chemically Induced Magnetic Polarization*, Wiley, New York (1973).
16. G. L. Closs and L. E. Closs, *J. Amer. Chem. Soc.*, **91**, 4550 (1969).
17. R. G. Lawler, *Accounts Chem. Res.*, **5**, 25 (1972).
18. G. Closs and A. D. Trifunac, *J. Amer. Chem. Soc.*, **92**, 2186 (1970); C. Walling and A. R. Lepley, *ibid.*, **93**, 546 (1971).
19. R. Kaptein, *J. Amer. Chem. Soc.*, **94**, 6251 (1972).
20. H. D. Roth and M. L. Manion, *J. Amer. Chem. Soc.*, **97**, 6886 (1975).
21. J. K. S. Wan, S.-K. Wong and D. A. Hutchinson, *Accounts Chem. Res.*, **7**, 58 (1974).
22. C. Lagercrantz, *J. Phys. Chem.*, **75**, 3466 (1971).
23. F. Nerdel and W. Brodowski, *Chem. Ber.*, **101**, 1398 (1968).
24. R. E. Kellogg and W. T. Simpson, *J. Amer. Chem. Soc.*, **87**, 4230 (1965).
25. P. S. Engel and B. M. Monroe, *Adv. Photochem.*, **8**, 245 (1971).
26. R. S. H. Liu and J. R. Edman, *J. Amer. Chem. Soc.*, **90**, 213 (1968).
27. N. C. Yang and S. P. Elliott, *J. Amer. Chem. Soc.*, **90**, 4194 (1968).
28. W. G. Herkstroeter and G. S. Hammond, *J. Amer. Chem. Soc.*, **88**, 4769 (1966).
29. J. G. Calvert and J. N. Pitts, *Photochemistry*, Wiley, London (1966), p. 728; B. Muel and C. Malpiece, *Photochem. Photobiol.*, **10**, 283 (1969); S. L. Murov, *Handbook of Photochemistry*, Dekker, New York (1973).
30. H. A. Taylor in J. M. Fitzgerald (ed.), *Analytical Photochemistry and Photochemical Analysis*, Marcel Dekker, New York (1971), p. 91.
31. J. G. Calvert and J. N. Pitts, *Photochemistry*, Wiley, London (1966), p. 780; P. de Mayo and H. Shizuka in W. R. Ware (ed.), *Creation and Detection of the Excited State*, Volume 4, Dekker, New York (1976).
32. M. D. Shetlar, *Mol. Photochem.*, **5**, 287 (1973).
33. F. G. Moses, R. S. H. Liu and B. M. Monroe, *Mol. Photochem.*, **1**, 245 (1969).
34. J. C. Dalton and N. J. Turro, *Mol. Photochem.*, **2**, 133 (1970); P. J. Wagner in A. A. Lamola (ed.), *Creation and Detection of the Excited State*, Volume 1, part A, Marcel Dekker, New York (1971), p. 173.
35. J. E. Gano and N. A. Marron, *Mol. Photochem.*, **8**, 141 (1977) and references therein.
36. J. B. Birks (ed.), *Organic Molecular Photophysics*, Volume 1, Wiley, London (1973), chapter 8.
37. P. J. Wagner, *J. Amer. Chem. Soc.*, **89**, 5715 (1967).
38. A. A. Lamola and G. S. Hammond, *J. Chem. Phys.*, **43**, 2129 (1965); J. B. Birks (ed.), *Organic Molecular Photophysics*, Volume 2, Wiley, London (1975), chapter 3.
39. K. Salisbury, *Tetrahedron Letters*, 737 (1971).
40. B. K. Dunning, D. H. Shaw and H. O. Pritchard, *J. Phys. Chem.*, **75**, 580 (1971).
41. C. A. Parker and T. A. Joyce, *Trans. Faraday Soc.*, **65**, 2823 (1969).

Chapter 6

Orbital Symmetry and Photochemistry

6.1 INTRODUCTION

It is assumed that the reader already has some acquantance with orbital symmetry considerations, i.e. the Woodward–Hoffmann rules,[1] so that the consideration of this topic here will be highly condensed. The Woodward–Hoffmann rules apply only to *pericyclic* reactions, that is to reactions occurring in a concerted manner *via* cyclic transition states. The distinction between concerted and non-concerted reactions is this: in concerted reactions the bond-forming and bond-breaking processes occur simultaneously (in concert), so that in one step the reactants are transformed into products. Such a situation can be represented by an energy profile such as Figure 6.1.

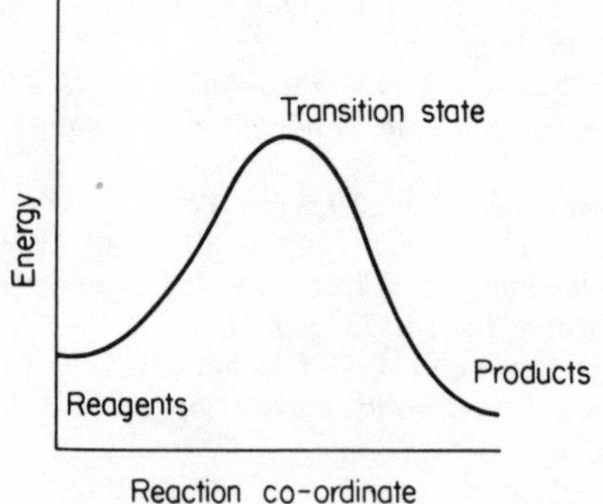

Figure 6.1. Energy profile for a concerted reaction

By contrast, in non-concerted reactions bond-forming and bond-breaking processes occur consecutively, so that intermediates are formed. The energy profile for such a reaction will have a minimum or minima corresponding to the various intermediates (see Figure 6.2).

The dimerization of two ethylene molecules to form cyclobutane or the reverse cleavage of cyclobutane into two ethylene molecules can be taken as an illustration. In the concerted process the formation or breaking of one σ-bond

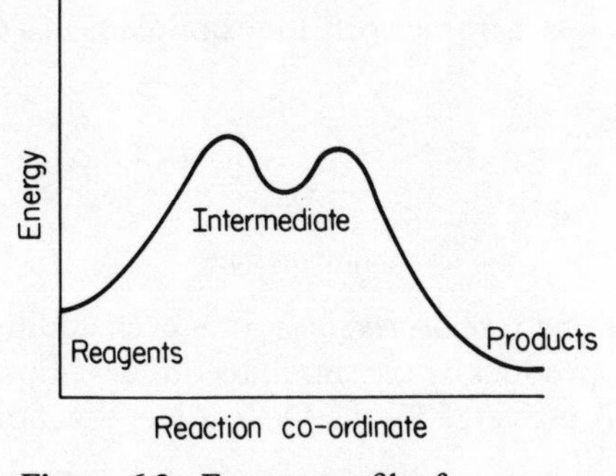

Figure 6.2. Energy profile for a non-concerted reaction

is controlled by, and coupled with, the formation or breaking of the other *via* a cyclic transition state, while a non-concerted pathway involves a biradical intermediate (Figure 6.3).

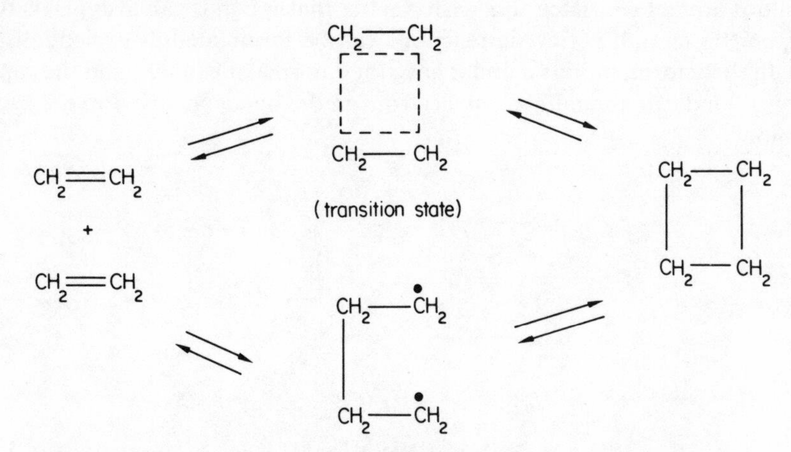

Figure 6.3. Concerted and non-concerted pathways for ethylene dimerization

6.2 CLASSES OF PERICYCLIC REACTION

There are three main classes of pericyclic reaction:

(i) *Electrocyclic Reactions.* These are processes in which a σ-bond develops between two atoms of a conjugated π-system to give a cyclic system with one fewer double bond. The reverse ring-opening is also regarded as an electrocyclic process. An example is the interconversion of buta-1,3-diene and cyclobutene (6.1).

$$\rightleftarrows \quad \text{transition state} \quad \rightleftarrows \qquad (6.1)$$

(ii) *Sigmatropic Reactions.* In these processes a σ-bond migrates to a new position with respect to a π-framework, for example as in a Cope rearrangement (6.2).

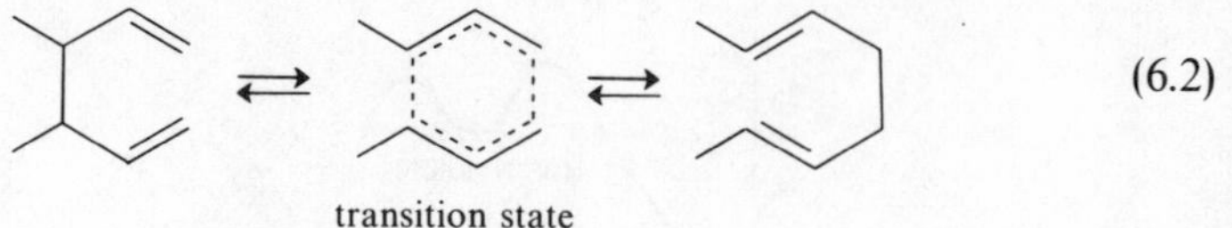

(iii) *Cycloadditions and Cycloreversions.* In a cycloaddition two independent systems form a cyclic product by the simultaneous development of two σ-bonds. A classical example is provided by the Diels–Alder reaction (6.3).

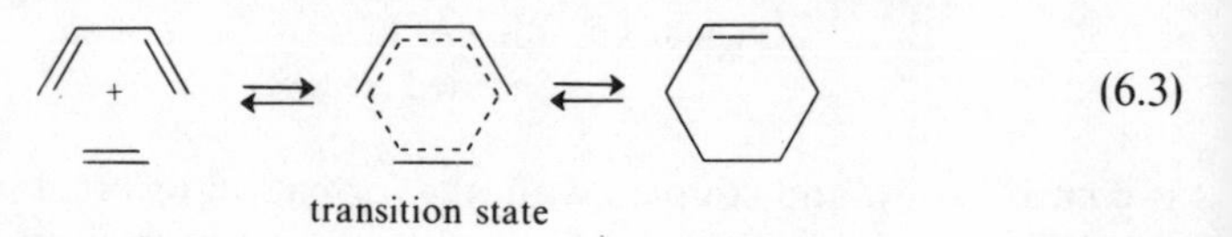

Cycloreversion is the inverse process. It is important to note that cycloadditions are not restricted to π-systems, but that σ-bonds can also participate (6.4), and in fact all pericyclic reactions can be formulated as cycloadditions (6.5). In these formulations σ and π have their normal meanings, and the superscripts (2) indicate the number of electrons in the bonds involved in the cycloaddition.

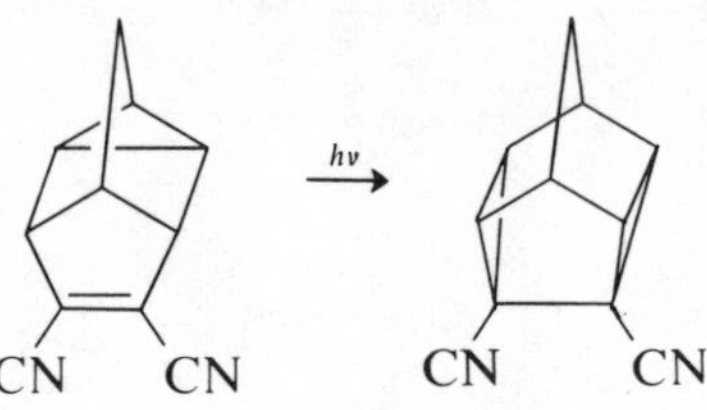

Formally

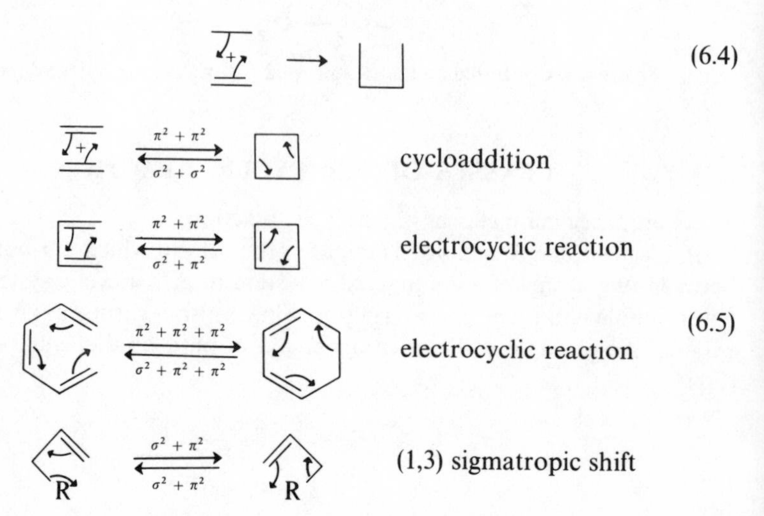

6.2.1 Stereochemical Aspects

The whole point of the Woodward–Hoffmann rules is to relate the geometrical changes occurring in a pericyclic reaction to the electronic wavefunctions or orbitals of the starting material(s). Thus for electrocyclizations, two modes of geometrical change must be distinguished—conrotatory (6.6) and disrotatory (6.7) ring-closures.

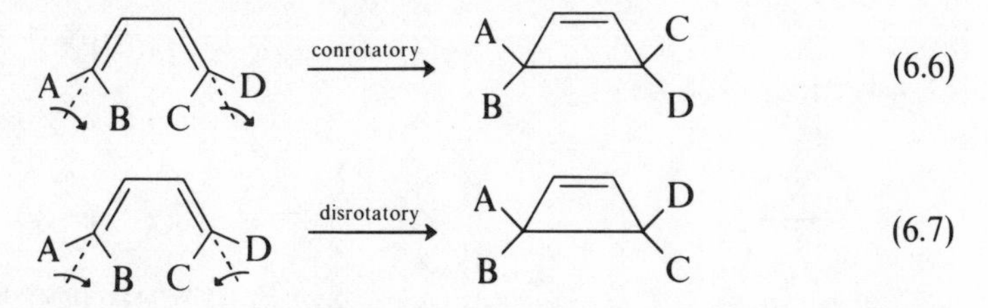

For sigmatropic reactions the concern is with whether the migrating group moves *suprafacially* (in which case it remains on the same face of the molecule) or *antarafacially* (in which case it migrates to the opposite face of the molecule), and with whether the migrating group retains its stereochemistry or is inverted. Equation (6.8) illustrates a (1, 3) sigmatropic shift which occurs suprafacially and with inversion of configuration in the migrating group.

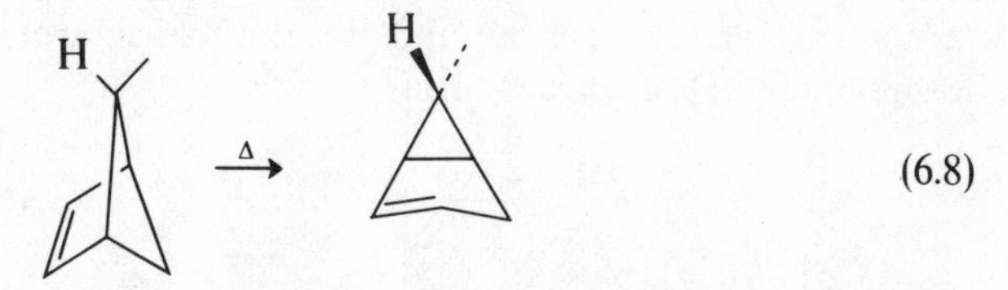

In cycloadditions involving π-systems *cis*- and *trans*-modes of addition to the double bonds can be easily differentiated (6.9), but with cycloadditions involving σ-bonds the question also arises as to whether the addition occurs with inversion or retention of configuration of the tetrahedral carbon atoms at the termini of the σ-bond.

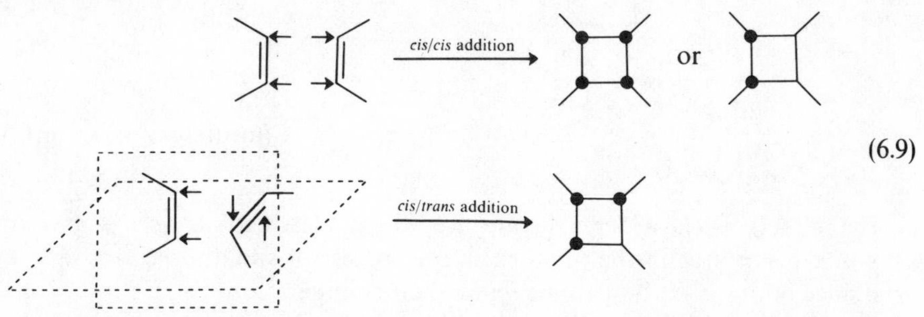

One way of bringing all these stereochemical features into a single system of nomenclature is to regard all pericyclic reactions as cycloadditions (as outlined above) and to generalize the concepts of antarafacial and suprafacial modes of addition (Figure 6.4).

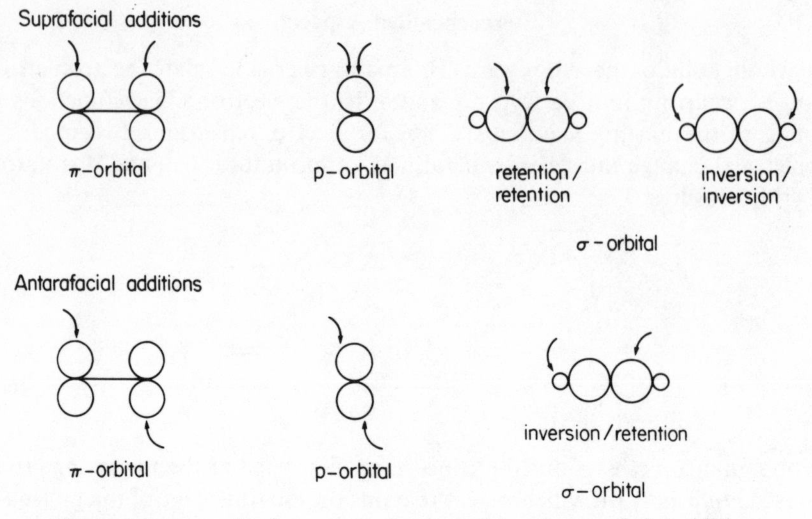

Figure 6.4. Generalized concepts of suprafacial and antarafacial additions

Using the abbreviations 's' and 'a' for suprafacial and antarafacial, pericyclic reactions can now be designated precisely. Thus the photochemical disrotatory electrocyclization of conjugated dienes can be formulated (6.10) as either $(\pi^2s + \pi^2s)$ or $(\pi^2a + \pi^2a)$, and the thermal conrotatory cyclization can be formulated (6.11) as $(\pi^2a + \pi^2s)$.

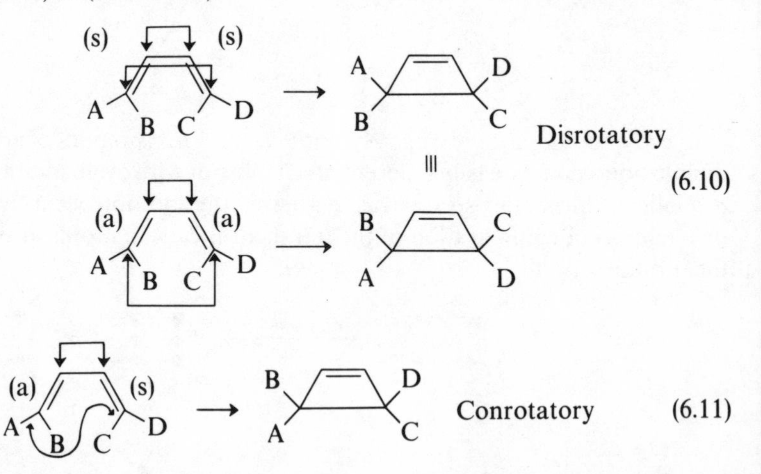

The *cis–cis* Diels–Alder reaction would be classified as $(\pi^4s + \pi^2s)$ or $(\pi^2s + \pi^2s + \pi^2s)$, and the photochemical process of equation (6.4) is $(\pi^2s + \sigma^2s)$ since both sp^3-carbon atoms retain their configuration.

6.3 THEORETICAL TREATMENTS

There are a number of formally different, but theoretically related, treatments of pericyclic reactions, *inter alia* the phase continuity principle,[2] frontier molecular orbitals,[3] the aromatic transition state concept,[4] correlation diagrams,[5] and the Woodward–Hoffmann general rule.[6] In this book only the last two treatments are considered.

6.3.1 Correlation Diagrams

This method, though powerful, is restricted in its application to symmetrical systems reacting *via* symmetrical transition states. The relevant orbitals† of the starting materials and products are ordered in increasing energy and classified as symmetric (*S*) or antisymmetric (*A*) with respect to the symmetry operations (reflections, rotations etc.) corresponding to the symmetry elements of the transition state. They are then correlated appropriately. The procedure may be illustrated for the electrocyclic reaction butadiene ⇄ cyclobutene. Note that the transition states for disrotatory and conrotatory cyclizations have a mirror plane and a twofold rotation axis respectively (Figure 6.5). The orbital correlation diagram takes the form of Figure 6.6.

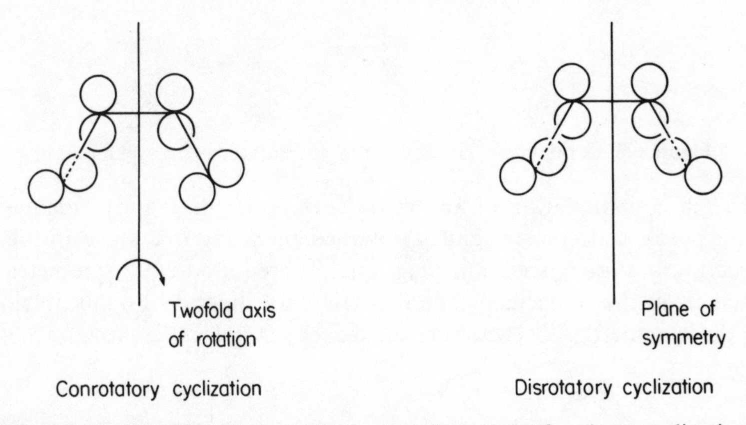

Figure 6.5. Symmetry elements in the transition states for electrocyclization

It can be readily seen that in the conrotatory mode, ground state butadiene ($\psi_1^2\psi_2^2$, i.e. doubly occupied ψ_1 and ψ_2) correlates with ground state cyclobutene ($\sigma^2\pi^2$), and that in the disrotatory mode, ground state butadiene correlates with *doubly* excited cyclobutene ($\sigma^2\pi^{*2}$) and ground state cyclobutene with doubly excited butadiene ($\psi_1^2\psi_3^2$). From this it may be concluded on energy grounds that the thermal electrocyclization butadiene ⇄ cyclobutene is allowed if conrotatory and forbidden if disrotatory. With respect to excited states, it is apparent that S_1-butadiene ($\psi_1^2\psi_2\psi_3$) correlates with S_1-cyclobutene ($\sigma^2\pi\pi^*$) only in the disrotatory mode, leading to the suggestion that photochemical interconversion of butadiene and cyclobutene should occur in the disrotatory fashion (see p. 171, however).

† i.e. those orbitals actually involved in the bonding changes in the reaction.

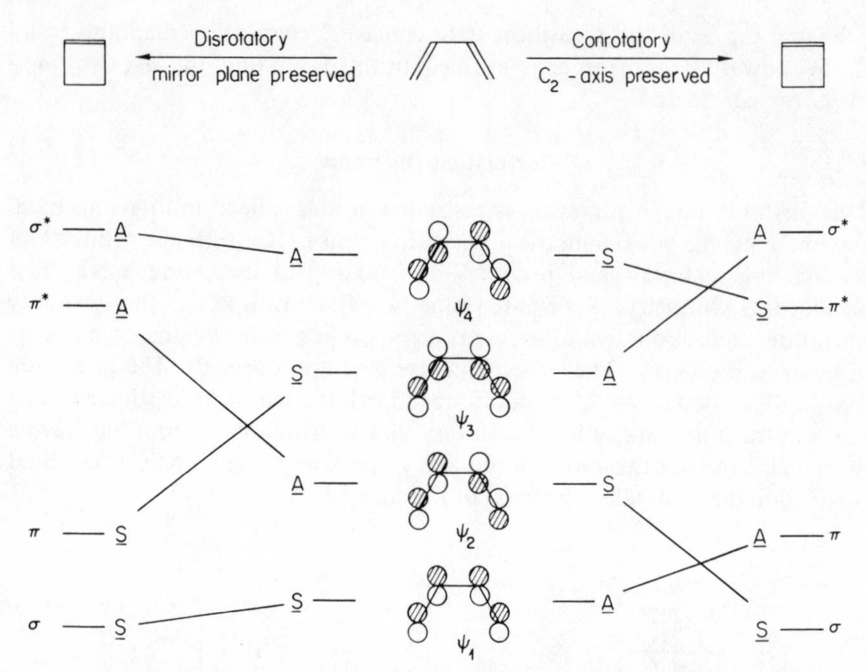

Figure 6.6. Orbital correlation diagram for butadiene ⇄ cyclobutene

Although contemplation of an *orbital* correlation diagram is adequate for most purposes, a deeper insight is obtained by going one stage further and constructing a *state* correlation diagram. The required state symmetries are obtained from the molecular orbital correlation diagram by multiplying together the 'symmetries' of each electron in each orbital, using the rules $S \times S =$

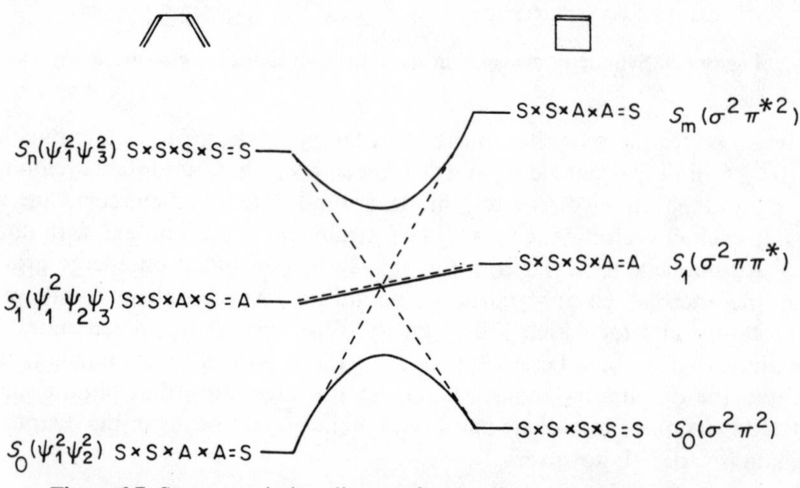

Figure 6.7. State correlation diagram for the disrotatory cyclization of butadiene

$A \times A = S$, and $S \times A = A$. By these means the state correlation diagram shown in Figure 6.7 is obtained for the disrotatory cyclization of butadiene.

Since, according to the *orbital* diagram (Figure 6.6), the ψ_1 and ψ_2 orbitals of butadiene transform into σ and π^* orbitals of cyclobutene, it follows that S_0-butadiene is correlated with a highly excited state S_m of cyclobutene. This and other correlations are exhibited on the state diagram (Figure 6.7) by dotted lines. Because of the non-crossing rule (that lines connecting states of like symmetry may not cross), the actual correlations are those given in Figure 6.7 by the full lines. Hence, since in a state correlation diagram one is plotting some ill-defined reaction co-ordinate as horizontal axis against energy, it follows that the disrotatory interconversion of butadiene and cyclobutene is therefore forbidden. On the other hand, no such barrier is apparent in the S_1-state. The state correlation diagram therefore provides a crude representation of a particular cross-section of the relevant multi-dimensional energy surfaces.

A minor technical difficulty arises in some reactions when the orbitals involved are apparently neither symmetric nor antisymmetric with respect to a particular symmetry operation. This is the case for the σ-orbitals of cyclohexene formed in the Diels–Alder reaction (Figure 6.8).

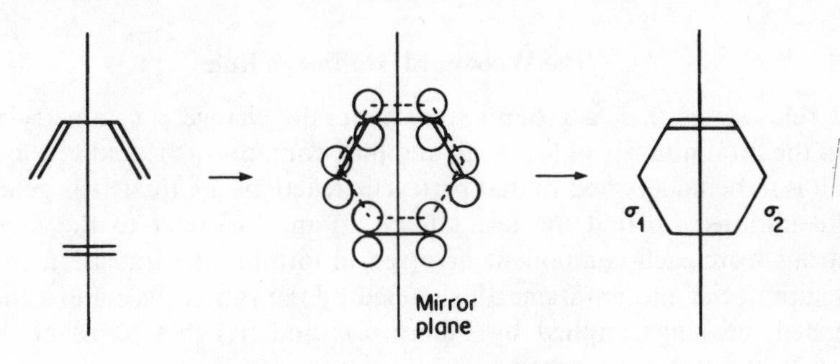

Figure 6.8. The Diels–Alder reaction

Reflection of σ_1 in the symmetry plane converts it into σ_2 and not into σ_1 or $-\sigma_1$. This sort of problem is overcome by mixing σ_1 and σ_2 to form multicentre orbitals $(\sigma_1 + \sigma_2)$ and $(\sigma_1 - \sigma_2)$, respectively symmetric and antisymmetric to the mirror plane (Figure 6.9).

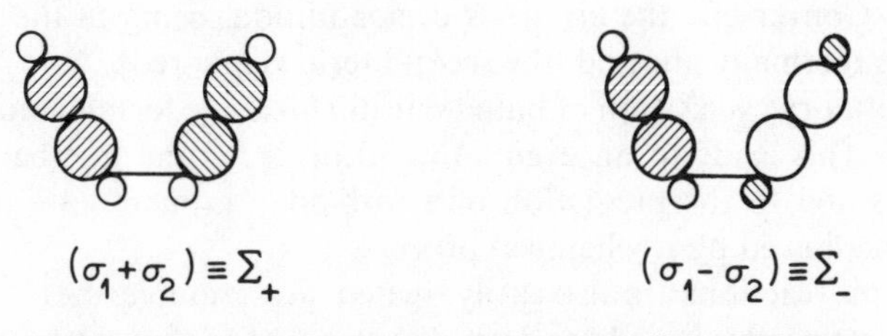

Figure 6.9. Symmetric and antisymmetric molecular orbitals of the σ-bonds formed in the Diels–Alder reaction between butadiene and ethylene

This device then permits the construction of the orbital correlation diagram of Figure 6.10, which shows that the Diels–Alder addition is allowed in the ground state.

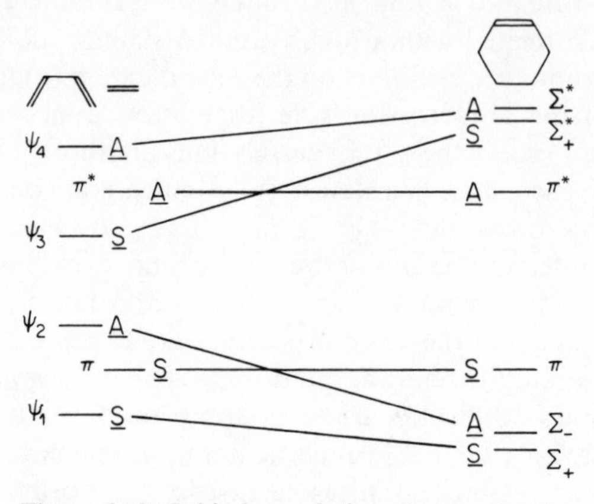

Figure 6.10. Orbital correlation diagram for the Diels–Alder cycloaddition

6.3.2 The Woodward–Hoffmann Rule

This rules states that 'a ground state pericyclic change is symmetry-allowed when the total number of $(4q + 2)_s$ and $(4r)_a$ components is odd'. In using this rule it is to be understood (i) that pericyclic reactions are treated as generalized cycloadditions; (ii) that the terms $(4q + 2)$ and $(4r)$ refer to the number of electrons from each component involved in forming the transition state; (iii) that suprafacial and antarafacial, signified by the subscripts s and a, have the extended meanings implied by Figure 6.4; and (iv) that terms of the type $(4q + 2)_a$ and $(4r)_s$ are ignored.

A few examples will clarify the procedure. Consider the dimerization of olefins (6.9). The *cis–cis* cyclodimerization takes the form $(\pi^2\text{s} + \pi^2\text{s})$, corresponding to two $(4q + 2)_s$ components ($q = 0$). The number of relevant components is even, and the reaction will be forbidden in the ground state. Since, for reasons to be discussed later, reactions thermally forbidden are allowed photochemically, we can conclude that this mode of cycloaddition will occur on irradiation. Conversely, the *cis–trans* cycloaddition, being of the type $(\pi^2\text{s} + \pi^2\text{a})$, will be thermally allowed (the second term is ignored).

The disrotatory cyclization of butadiene (6.11) can be formulated as $\pi^4\text{s}$ or as $(\pi^2\text{s} + \pi^2\text{s})$. This leads to an even value (0 or 2) for the number of relevant components and to the prediction of a forbidden ground state reaction and therefore an allowed photochemical process.

Sigmatropic reactions are also easily treated. For example, the (1, 3) migration of an alkyl group suprafacially and with inversion of configuration (see equation 6.8) is a $(\pi^2\text{s} + \sigma^2\text{a})$ process (Figure 6.11). The number of relevant components is odd, and the reaction is allowed in the ground state.

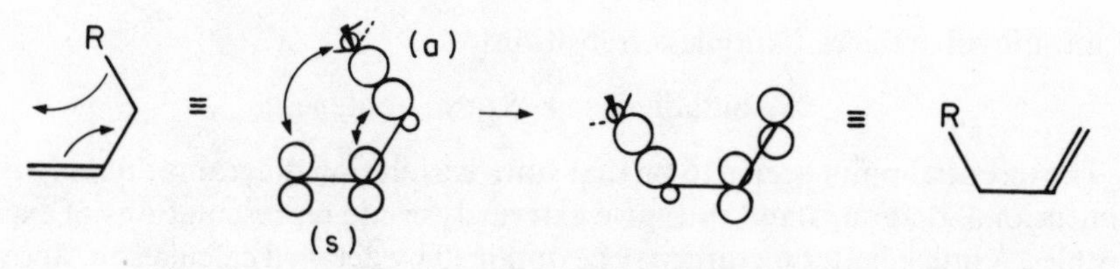

Figure 6.11. A suprafacial (1,3)-sigmatropic migration with inversion of the migrating group

6.4 ORBITAL SYMMETRY AND PHOTOCHEMICAL REACTIONS

Extensive investigations have shown that pericyclic reactions proceed in opposite senses in the ground and excited states (6.12).

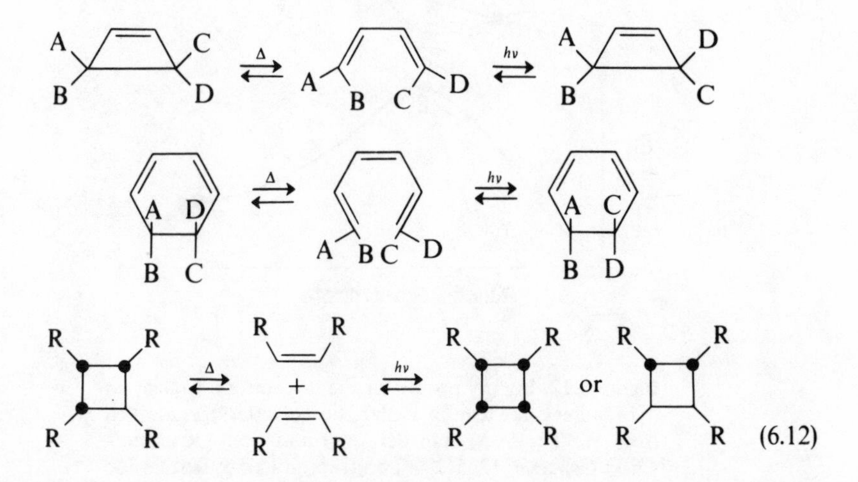

In order to understand the origin of this dichotomy between thermal and photochemical reactions, and in order to illustrate the nature of the problem, the electrocyclization butadiene ⇄ cyclobutene will be reconsidered. From an examination of the state correlation diagram (Figure 6.7) it is seen that in the disrotatory mode the ground states do not correlate and hence are separated by an energy barrier, while the S_1 states do correlate. It might therefore be thought that herein lies the explanation of why the disrotatory cyclization of butadiene occurs photochemically but not thermally. This, however, would be a delusion, for although the disrotatory conversion of S_1-butadiene into S_1-cyclobutene is allowed by orbital symmetry, it cannot possibly occur because S_1-cyclobutene is about 200–250 kJ mol^{-1} (50–60 kcal mol^{-1}) more endothermic than S_1-butadiene. It follows that the reaction sequence cannot be

$$S_0\text{-butadiene} \rightarrow S_1\text{-butadiene} \rightarrow S_1\text{-cyclobutene} \rightarrow S_0\text{-cyclobutene}$$

It must involve the radiationless transition†

$$S_1\text{-butadiene} \rightsquigarrow S_0\text{-cyclobutene}$$

The essential point seems to be that state correlation diagrams are only two-dimensional diagrams and thus give extremely crude representations of energy profiles. A much better picture may be obtained by detailed calculation. According to valence-bond calculations,[7] the energy profiles for the disrotatory cyclization of butadiene actually have the form of Figure 6.12, and all now becomes clear.

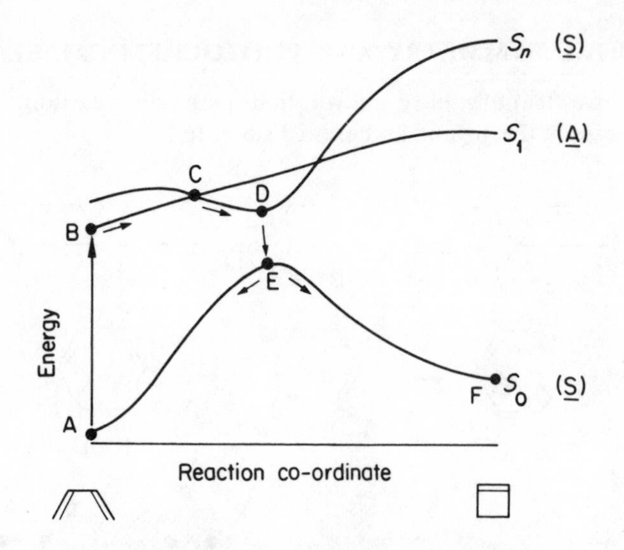

Figure 6.12. Energy profile for the disrotatory cyclization of butadiene (see text for explanation of lettering) (adapted from W. Th. A. M. van der Lugt and L. J. Oosterhoff, *Chem. Commun.*, 1235 (1968), and reproduced by permission of the Chemical Society)

S_0-Butadiene (A) is excited to the S_1-state (B), and molecular vibrations enable it to progress up the slight energy hill to the point (C) where it can cross to the symmetric S_n-state and coast downhill to the energy minimum (D). Here the molecule can undergo rapid radiationless transition (small energy gap) to S_0 (E), where it has the choice of returning to butadiene or proceeding to (F) and forming cyclobutene. In the conrotatory mode (Figure 6.13) such a sequence is impossible.

The above considerations have very general implications. Referring to Figure 6.7, the energy barrier which makes the ground state reaction forbidden

† It should be noted that apart from proton transfer reactions $MH^* \rightleftarrows M^* + H^+$, which give excited state products and therefore occur adiabatically on the S_1-energy surface, the preponderant majority of photochemical reactions involve radiationless transitions so that the excited starting material goes directly to ground state products (and *not* via the excited state of the products).

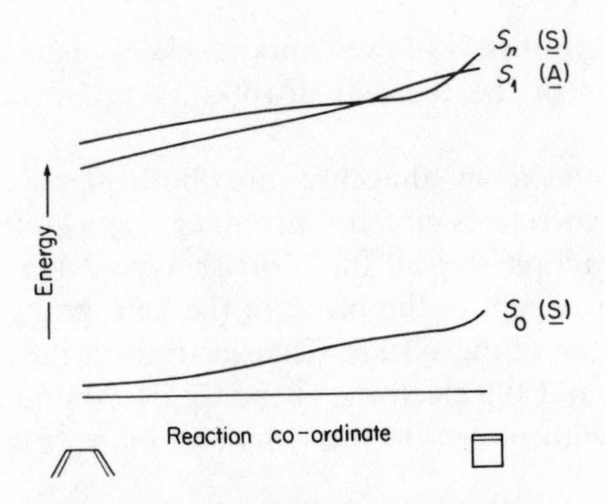

Figure 6.13. Energy profile for the conrotatory cyclization of butadiene (adapted from W. Th. A. M. van der Lugt and L. J. Oosterhoff, *Chem. Commun.*, 1235 (1968), and reproduced by permission of the Chemical Society)

arises because the two ground states are correlated with upper excited states. Inevitably there must be a corresponding minimum in the energy surface of an upper state of the same symmetry. Since upper states are densely packed, it will normally be the case that the upper state with the minimum will cross lower excited states and probably also S_1. Therefore correlation diagrams for reactions forbidden in the ground state will usually be of the form of Figure 6.14 (for example, see Figure 6.12) and not that of Figure 6.15.

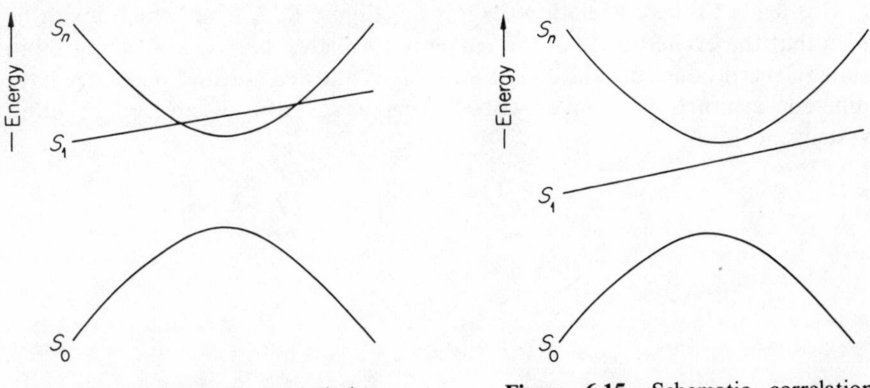

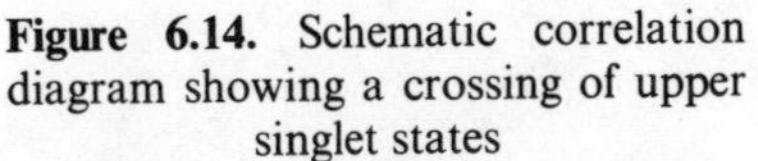

Figure 6.14. Schematic correlation diagram showing a crossing of upper singlet states

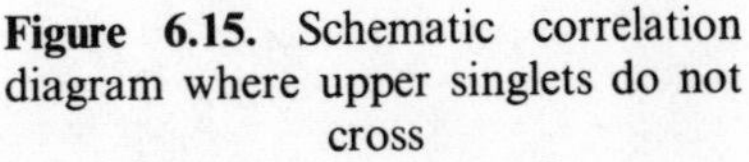

Figure 6.15. Schematic correlation diagram where upper singlets do not cross

From this it follows that, in general, when orbital symmetry forbids a reaction in the ground state, that same reaction will be allowed photochemically. It should always be borne in mind that reactions allowed by orbital symmetry do not necessarily occur—there may, for example, be insuperable geometric restraints, as in the symmetry-allowed antarafacial sigmatropic rearrangements in cyclic systems or in the thermal disrotatory cyclization of butadiene to *trans*-cyclobutene.

It has recently been shown[8] that there are photochemical reactions in which the excited reagent correlates *directly* with the ground state product. Consider the hydrogen abstraction step in the Norrish type 2 reaction. It is assumed that the abstraction occurs in the plane of the keto group, which is taken to be the symmetry plane of the system. The electrons of the n and π^* orbitals of the carbonyl group and the electrons of the C—H σ-bond are considered and classified as σ or π with respect to the symmetry plane (Figure 6.16).

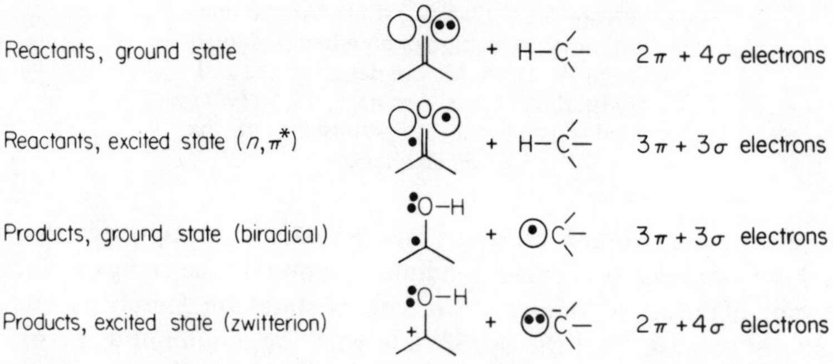

Reactants, ground state	$2\pi + 4\sigma$ electrons
Reactants, excited state (n,π^*)	$3\pi + 3\sigma$ electrons
Products, ground state (biradical)	$3\pi + 3\sigma$ electrons
Products, excited state (zwitterion)	$2\pi + 4\sigma$ electrons

Figure 6.16. Classification of states in the reactants and products for hydrogen abstraction

This leads to the correlation diagram of Figure 6.17, from which it can be seen that the excited state of the reagents correlates *directly* with the ground state of the product biradical. The crossing is allowed because the states have different symmetries. Similar correlations have been discovered in other systems.

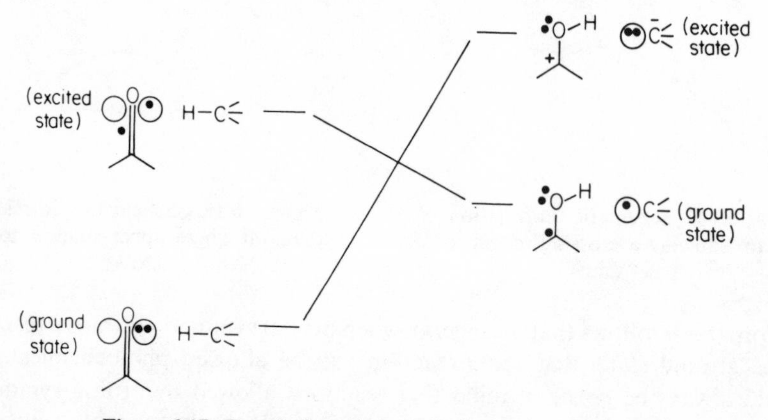

Figure 6.17. Correlation diagram for hydrogen abstraction

REFERENCES

1. An excellent introduction is T. L. Gilchrist and R. C. Storr, *Organic Reactions and Orbital Symmetry*, Cambridge University Press, London (1972).
2. W. A. Goddard, *J. Amer. Chem. Soc.*, **94**, 793 (1972).
3. Reviewed by K. Fukui and H. Fujimoto in B. S. Thyagarajan (ed.), *Mechanisms of Molecular Migration*, volume 2, Interscience, London (1969), p. 117.
4. M. J. S. Dewar, *Molecular Orbital Theory of Organic Chemistry*, McGraw-Hill, New York (1969); H. E. Zimmerman, *J. Amer. Chem. Soc.*, **88**, 1564 and 1566 (1966); H. E. Zimmerman, *Angew. Chem. Intern. Ed.*, **8**, 1 (1969).
5. H. C. Longuet-Higgins and E. W. Abrahamson, *J. Amer. Chem. Soc.*, **87**, 2045 (1965).
6. R. B. Woodward and R. Hoffmann, *The Conservation of Orbital Symmetry*, Verlag Chemie, Weinheim (1970).
7. W. Th. A. M. van der Lugt and L. J. Oosterhoff, *Chem. Commun.*, 1235 (1968).
8. L. Salem, W. G. Dauben and N. J. Turro, *J. Chim. Phys.*, **70**, 694 (1973); L. Salem, *J. Amer. Chem. Soc.*, **96**, 3486 (1974).

Problems

The problems cover most of the major subject areas discussed in the text, and the sequence of problems corresponds approximately to the order of presentation of material in the text. Numerical answers and brief comments are provided at the end of the section.

1. What is the energy (kJ mol^{-1} or kcal mol^{-1}) associated with ultraviolet radiation of wavelength
   (i) 184·7 nm, (ii) 253·7 nm, (iii) 366·0 nm?

2. A medium-pressure mercury arc lamp, used in conjunction with a chemical filter solution, provides 6 watts of energy at 313·0 nm which is completely absorbed by a sample surrounding the lamp. How long will it take for 0·1 mole of product to be formed if the quantum yield for product formation is 0·3?

3. Furan has a long-wavelength absorption band with $\lambda_{max} \sim 250$ nm. When light of this wavelength is passed through a 10 cm cell containing a 10^{-2} M solution of furan, 20·6% of the incident light is absorbed.
   (i) What is the optical density ($\log I_0/I$) of the solution?
   (ii) What percentage of the incident light would be transmitted by a 20 cm cell of the same solution?
   (iii) What is the molar extinction coefficient of furan at 250 nm?
   (iv) Is the absorption likely to arise from a $\pi \longrightarrow \pi^*$ transition?

4. The absorption spectrum of *p*-benzoquinone is given in Figure 2.20 on p. 34. Estimate the oscillator strength of the long-wavelength transition by re-plotting the curve on a wave-number scale and finding the area under the curve (by 'counting squares' or by 'cutting out and weighing').

5. Many molecules in solution have absorption bands whose width at half-height (i.e. where $\varepsilon = \frac{1}{2}\varepsilon_{max}$) is ~ 5000 cm^{-1}. The band area may then be calculated approximately as that of a rectangle of length ε_{max} and breadth 5000 cm^{-1}. On this basis calculate the oscillator strength:
   (i) of a transition giving rise to a band with $\varepsilon_{max} = 10^3$ l mol^{-1} cm^{-1};
   (ii) of the benzene ${}^1A_{1g} \longrightarrow {}^1B_{2u}$ transition at 256 nm, for which $\varepsilon_{max} = 160$ l mol^{-1} cm^{-1} and the width at half-height is 4000 cm^{-1}.

6. Estimate the radiative lifetime of an excited state that decays to the ground state with the emission of 490 nm radiation. Assume that the transition is fully allowed. (*Hint:* equations (2.2) and (2.3) will help.)

$$c = 3{\cdot}0 \times 10^{10}\ \mathrm{cm\,s^{-1}}$$
$$e = 4{\cdot}8 \times 10^{-10}\ \mathrm{cm^{\frac{3}{2}}\,g^{\frac{1}{2}}\,s^{-1}}$$
$$m_e = 9{\cdot}1 \times 10^{-28}\ \mathrm{g}$$

7. A nitrogen laser produces pulses with a peak power of 200 kW. Assuming that the pulses are rectangular in shape and have a duration of 10 ns, how many photons ($\lambda = 337$ nm) are emitted in each pulse?

8. The wavelengths of the 0–0 transitions of a phenol and its anion in aqueous solution at 300 K are 475 and 550 nm respectively. If the $\mathrm{p}K_a$ of the phenol in its ground state is 9·2, what is the $\mathrm{p}K_a$ of its lowest excited singlet state?

9. Suggest a method of obtaining the $\mathrm{p}K_a$ value for a triplet state through observation of the triplet–triplet absorption of the acid and its conjugate base.

10. The $\Delta(\mathrm{p}K_a)$ values $[\Delta(\mathrm{p}K_a) = \mathrm{p}K_a(S_0) - \mathrm{p}K_a(S_1)]$ for *o*-, *m*- and *p*-methoxyphenol are 4·8, 7·0 and 4·6 respectively. Can you explain the substituent effect?

11. The long-wavelength absorption band of benzophenone is relatively weak ($\varepsilon_{max} \sim 100\ \mathrm{l\,mol^{-1}\,cm^{-1}}$) and the band is shifted to shorter wavelength when the solvent is changed from cyclohexane to ethanol. For 4,4′-bis-(dimethylamino)benzophenone the long-wavelength band is much stronger ($\varepsilon_{max} \sim 10\,000\ \mathrm{l\,mol^{-1}\,cm^{-1}}$) and there is a solvent effect in the opposite direction. Give an explanation of these observations.

12. The phosphorescence emission spectrum of benzophenone shows a vibrational progression with a separation of 1600–1700 $\mathrm{cm^{-1}}$ (C=O stretching), but the phosphorescence spectrum of 2-acetylnaphthalene resembles that of 1-chloronaphthalene and shows a series of vibrational bands separated by $\sim 1400\ \mathrm{cm^{-1}}$ (aromatic ring vibration). Can you account for this?

13. For toluene the quantum yield for fluorescence is 0·14 and that for intersystem crossing is 0·50, and the singlet lifetime is $3{\cdot}4 \times 10^{-8}$ s. What is (i) the rate constant for intersystem crossing, and (ii) the intrinsic radiative lifetime?

14. Fluorescence of naphthalene occurs with a quantum yield of 0·19 and a lifetime of 9.6×10^{-8} s. Fluorescence from the second excited singlet state (S_2) cannot be detected ($\phi < 10^{-4}$). Estimate a lower limit for the rate constant for internal conversion $S_2 \rightsquigarrow S_1$, and state the assumptions that you have made.

15. The emission spectra in the figure below were obtained for anthracene at 77 K in different solvents, (a) ethanol, (b) l-chloropropane, (c) l-bromopropane, (d) l-iodopropane. Explain the solvent effect that is occurring.

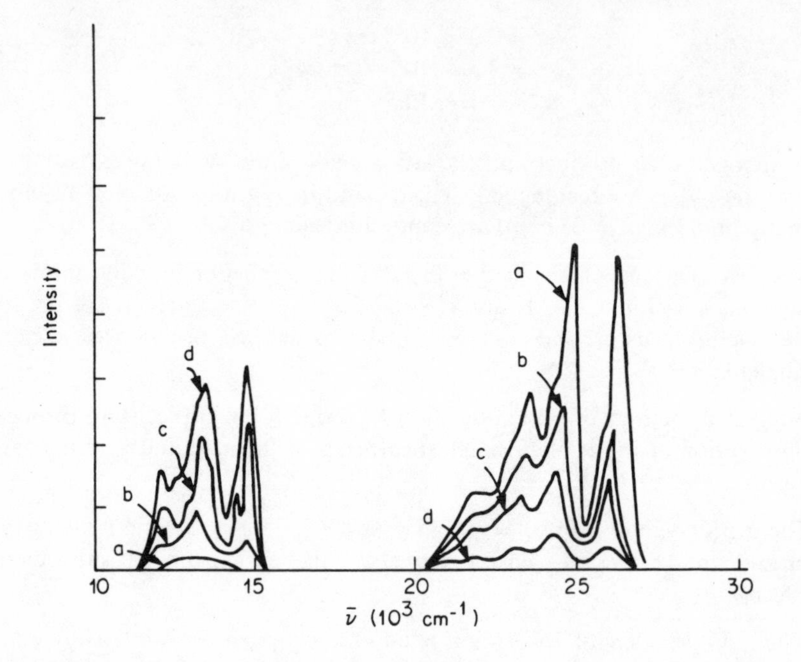

16. The figure below shows (a) the phosphorescence spectrum and (b) the delayed fluorescence excitation spectrum of anthracene. What can you tell from the spectra about the mechanism of delayed fluorescence?

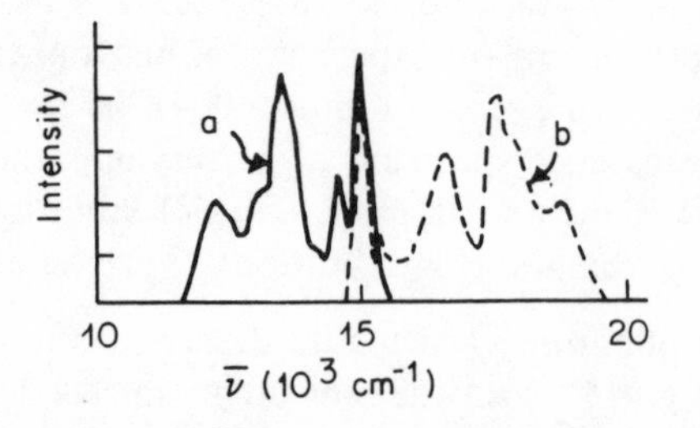

17. It is often observed that for fluorescent phenols whose anions are non-fluorescent the quantum yield for fluorescence is higher in D_2O than in H_2O. Can you suggest an explanation that accounts for this?

18. The spectra in the figure below are (a) the absorption and (b) the fluorescence excitation spectra of cyclobutanone taken in the gas phase. Why are the spectra not identical? Suggest experiments to confirm your hypothesis.

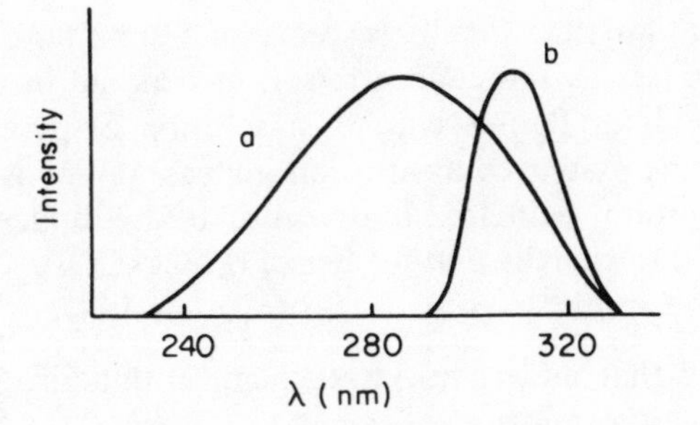

19. (i) Is spin–orbit coupling ('mixing') that involves the ground state and the first excited triplet state likely to be significant in most organic molecules? (ii) Other things being equal, is spin–orbit coupling likely to be greater for benzophenone ($E_T \sim 290$ (69), $E_S \sim 315$ (75) kJ (kcal) mol^{-1}) or for benzene ($E_T \sim 350$ (84), $E_S \sim 460$ (110) kJ (kcal) mol^{-1})? What will be a major consequence of this difference?

20. The following table records photophysical parameters relating to substituted naphthalenes at 77 K. Comment on the variations of the values with substituent.

1-$C_{10}H_7X$	ϕ_f	ϕ_p	τ_p (s)	k_{isc} (s^{-1})	k_f (s^{-1})
X = H	0·55	0·051	2·3	$\sim 10^6$	$\sim 10^6$
Me	0·85	0·044	2·1	$\sim 2 \times 10^5$	$\sim 3 \times 10^6$
OH	0·76	0·036	1·9	$\sim 2 \times 10^5$	$\sim 3 \times 10^6$
F	0·84	0·056	1·5	$\sim 2 \times 10^5$	$\sim 3 \times 10^6$
Cl	0·058	0·30	0·29	$\sim 1{\cdot}5 \times 10^7$	$\sim 3 \times 10^6$
Br	0·0016	0·27	0·002	$\sim 5 \times 10^8$	$\sim 3 \times 10^6$
I	<0·0005	0·38	0·002	$>3 \times 10^9$	$\sim 3 \times 10^6$

21. For biacetyl excited in the vapour phase with light of wavelength 436·5 nm, $\phi_{isc}(S_1 \rightsquigarrow T_1)$ is 0·97, ϕ_p is 0·15, τ_S (singlet lifetime) is 24 ns, and τ_T (triplet lifetime) is 0·0015 s. Calculate:
(i) the rate constant for intersystem crossing ($S_1 \rightsquigarrow T_1$);
(ii) the rate constant for fluorescence (assuming that the singlet state undergoes only fluorescence and intersystem crossing);
(iii) the rate constant for phosphorescence;
(iv) the rate constant for non-radiative decay of the triplet state.
(Reproduced by permission of the University of Southampton.)

22. At all concentrations of *o*-xylene the quantum yield for triplet formation (ϕ_T) is 0·28, but the triplet lifetime of *o*-xylene (estimated from the sensitized *cis–trans* isomerization of but-2-ene) decreases from 900 ns in 0·10 M solution to 17 ns in 3 M solution. Suggest a reason for this decrease in lifetime.

23. (i) The quantum yield for fluorescence of a solution of benzene in cyclohexane decreases as the concentration of benzene increases, whereas the quantum yield for fluorescence of a solution of pyrene in cyclohexane increases as the pyrene concentration increases. Why is this?
(ii) The quantum yield for fluorescence of 1,2-diphenylethane ($PhCH_2$-CH_2Ph) is 0·11, whereas that for benzil (PhCO.COPh) is only 10^{-3}. What is the reason for this?

24. A compound that undergoes photochemical dimerization is proposed to react by way of a singlet state exciplex:

$$A \xrightarrow{h\nu} {}^1A$$
$${}^1A \xrightarrow{k_{-1}} A$$
$${}^1A \xrightarrow{k_{isc}} {}^3A$$
$${}^1A + A \xrightarrow{k_e} E$$
$$E \xrightarrow{k_{-e}} 2A$$
$$E \xrightarrow{k_D} D$$

(A = monomer, E = exciplex, D = dimer).

(i) Derive an expression relating the quantum yield for dimer formation to the concentration of A.
(ii) What singlet state process has been neglected in the proposed scheme (n.b. k_{-1} and k_{-e} are composite rate constants covering both radiative and non-radiative decay)?

25. The triplet excimer of *o*-xylene (M) is very short-lived ($k_{isc} > 10^8\ s^{-1}$). Would the involvement of a 'double doublet' (A), in which a pair of radical ions is formed transiently, account for this?

$$[{}^3(M.M) \leftrightarrow {}^3(M^+.M^-) \leftrightarrow {}^3(M^-.M^+)] \rightarrow \underset{(A)}{({}^2M^+;{}^2M^-)} \rightarrow M + M$$

26. The figure shows (a) the absorption spectrum of vinylcarbazole (A), (b) the fluorescence spectrum of (A), and (c), (d), (e) the fluorescence spectrum of (A) in the presence of increasing concentrations of fumaronitrile (B). The concentration of (A) is $1{\cdot}79 \times 10^{-4}$ M, the solvent is benzene, and the fluorescence excitation wavelength 340 nm. Irradiation of the vinylcarbazole-fumaronitrile system at 340 nm has been shown to form poly-(vinylcarbazole) efficiently. Identify the emission bands, discuss their features, and suggest a possible mechanism to explain the observations.

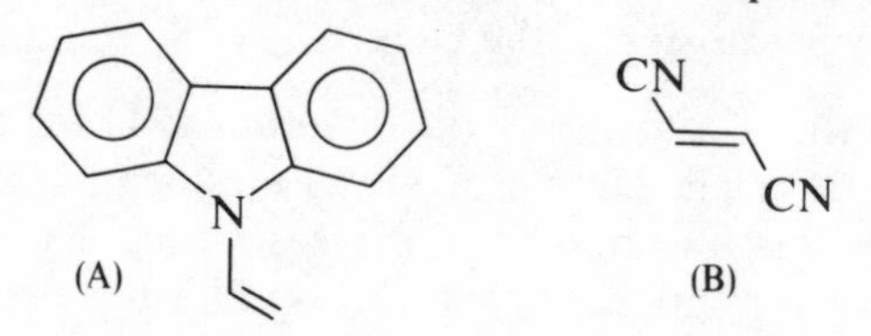

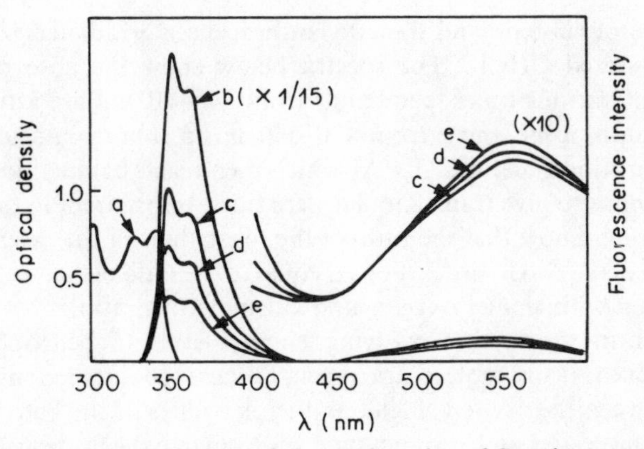

(Reproduced by permission of the University of Southampton.)

27. Excitation of naphthalene (D, 10^{-3} M) in the presence of 1,4-dicyanobenzene (A) leads to emission from the exciplex $(D^+.A^-)^*$ instead of naphthalene fluorescence. With A = 0·047 M, increasing D from 10^{-3} to 0·6 M quenches the exciplex emission and replaces it by a red-shifted structureless emission due to a new entity (X). The emission peaks of X and of the exciplex are red-shifted to a similar extent by increasing the solvent polarity. What inferences can you make concerning the nature of X?

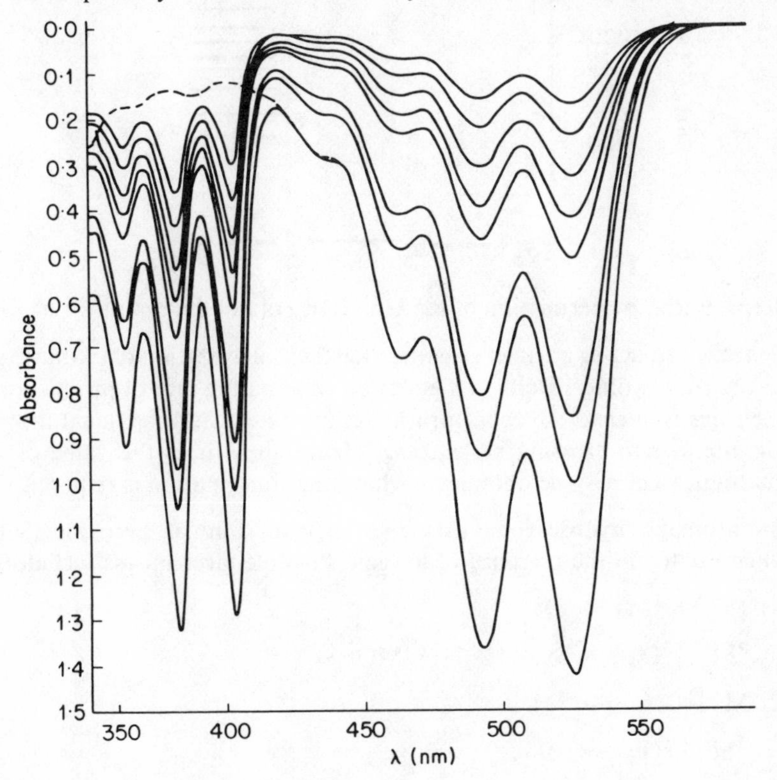

28. A mixture of rubrene and dimethylanthracene is irradiated ($\lambda = 546$ nm) in air-saturated $CHCl_3$. The spectra below show the absorption of the solution at various times; the bands from 340–410 nm are attributable to dimethylanthracene, those from 420–550 nm to rubrene. Assume that the light absorption generates $O_2(^1\Delta)$, which then reacts competitively with the hydrocarbons to give trans-annular peroxides, transparent in this region of the spectrum. Show that the ratio of the logarithms of the extent of photo-oxidation of the two hydrocarbons is equal to the ratio of their rate constants for reaction with singlet oxygen, and calculate this ratio.
29. The diagram shows the low-lying energy levels of benzophenone (B), naphthalene (N) and molecular oxygen. Discuss the interactions that might occur between the excited singlet and triplet states of the ketone or of the aromatic molecule and ground-state molecular oxygen, bearing in mind energetic considerations and the restrictions of the Wigner spin rules.

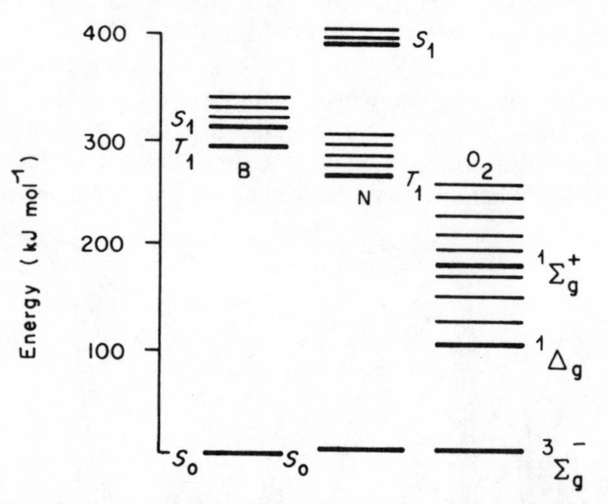

(Reproduced by permission of the University of Southampton.)

30. Benzene can act as a triplet sensitizer for the isomerization of *cis*-but-2-ene to the *trans*-isomer. Derive an equation relating the rate of production of the *trans*-isomer to the concentration of *cis*-butene and the quantum yield for intersystem crossing in benzene. Hence show how the value of this quantum yield may be obtained—what other information is required?

31. An aromatic hydrocarbon (M) gives a trans-annular peroxide (MO_2) when excited in the presence of oxygen. Possible mechanisms include:

(1) $M \xrightarrow{h\nu} {}^1M \rightarrow {}^3M$

$${}^3M + {}^3O_2 \rightarrow M + {}^1O_2 \rightarrow MO_2$$

(2) $M \xrightarrow{h\nu} {}^1M \rightarrow {}^3M$

$${}^3M + {}^3O_2 \rightarrow MO_2$$

Derive rate expressions for these two possibilities, and show how the mechanisms may be distinguished with their aid.

32. Irradiation of acenaphthylene (A) gives the *cis*- and *trans*- forms of the cyclobutane dimer (B). In the presence of oxygen, only the formation of the *trans*-dimer is strongly quenched. In the absence of oxygen the *cis*:*trans* ratio decreases markedly through the solvent sequence ethanol, l-chloropropane, l-bromopropane, l-iodopropane. Can you account for these observations?

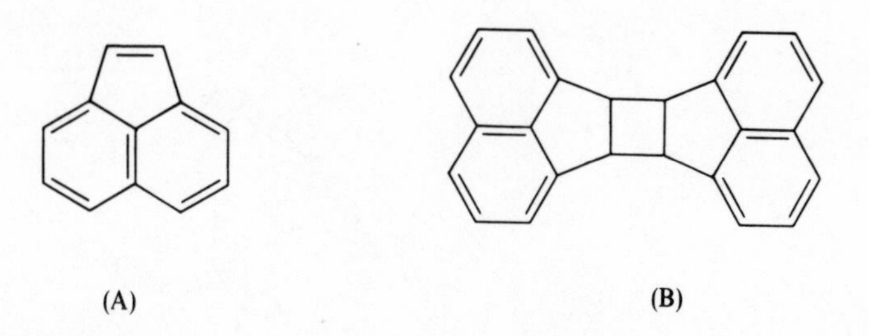

33. Irradiation of l-methylcyclopentene or l-methylcyclohexene in solution in *O*-deuteriomethanol gave the following results:

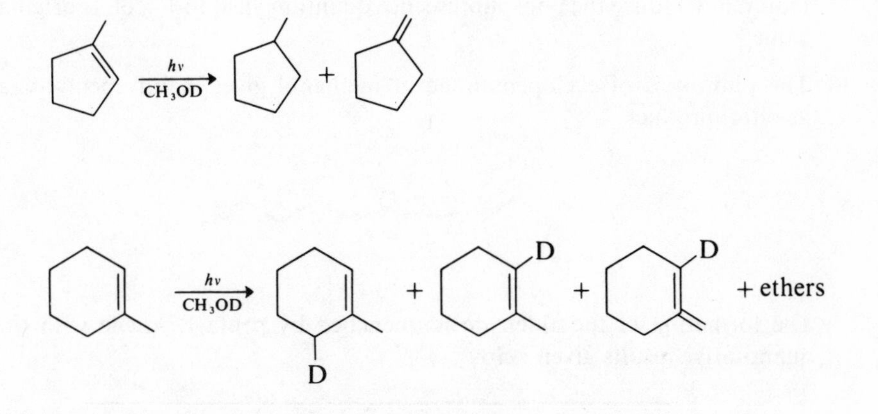

What mechanistic conclusions can you draw on the basis of this evidence alone?

34. Benzene and maleic anhydride react photochemically to give a 1:2 adduct (A). In the presence of fumaronitrile (*trans* NC—CH=CH—CN) no other adduct is formed, but with tetracyanoethylene, adduct (B) is produced, What mechanistic conclusions can you draw from the observations?

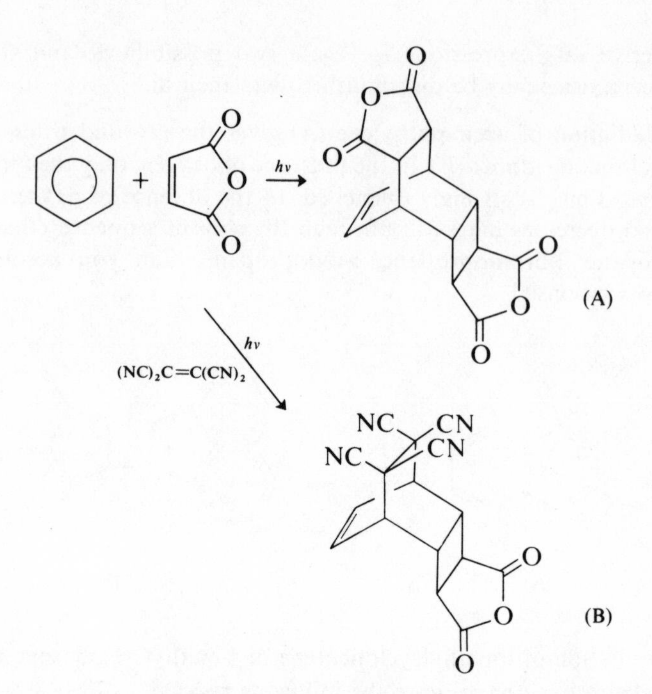

35. At room temperature biacetyl phosphoresces with a quantum yield of 0·23 and a triplet lifetime of 2×10^{-3} s. What is the concentration of quencher acting at the diffusion-controlled rate ($k_q = 10^{10}\,\mathrm{l\,mol^{-1}\,s^{-1}}$) that is required to reduce the phosphorescence quantum yield to 1% of its original value?

36. The photolysis of cyclopentanone in methanol gives mainly pent-4-enal as photoproduct:

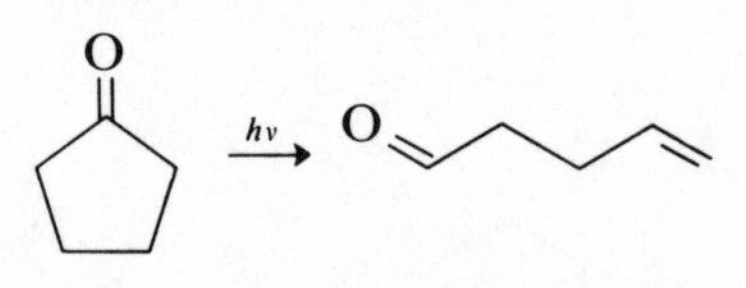

The formation of the aldehyde is quenched by penta-1,3-diene with the quantitative results given below:

[pentadiene] ($\mathrm{mol\,l^{-1}}$)	[aldehyde formed] (arbitrary units)
0·0000	24·8
0·0278	15·8
0·0556	13·0
0·112	7·84
0·167	5·75
0·278	3·35
0·556	2·1

What can you deduce about the mechanism from these results?

37. In a study of the quenching by penta-1,3-diene of the Norrish type 2 reaction of octan-2-one in benzene, the following results were obtained:

[penta-1,3-diene]	octan-2-one (% reacted)*
0·000	25·9
0·050	16·2
0·100	13·2
0·200	12·7
0·300	11·5
0·400	11·2
0·500	11·0
2·50	10·0

* Corrected for absorption of light by products.

What do these results tell you about the excited state(s) responsible for the reaction?

38. Two aliphatic ketones were irradiated in parallel in a merry-go-round apparatus using 313 nm radiation. One of them was hexan-2-one employed as a secondary actinometer (quantum yield for acetone production is 0·22), and by v.p.c. the amount of acetone produced in a given time was found to be 0·020 mol l^{-1}. The unknown ketone gave a product which was estimated to be formed in a yield of 0·0086 mol l^{-1} in the same period of time. What is the quantum yield for formation of this product, and what assumptions do you have to make in deriving an answer?

39. Classify the following reactions using the symbols associated with generalized cycloadditions [$(\pi^2 + \pi^2)$, $(\pi^2s + \sigma^2a)$, etc.]:

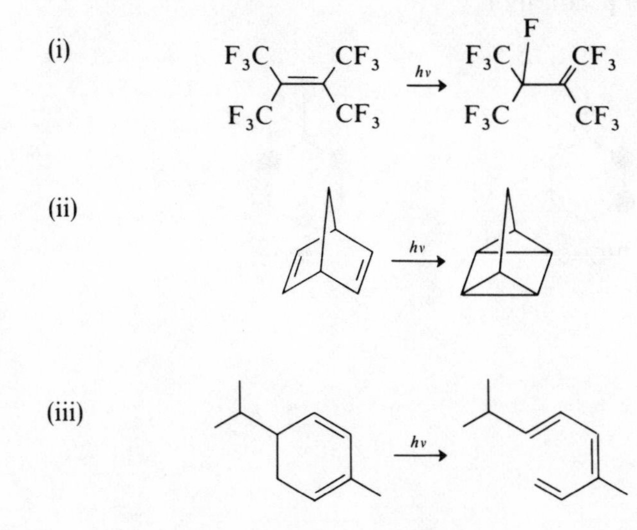

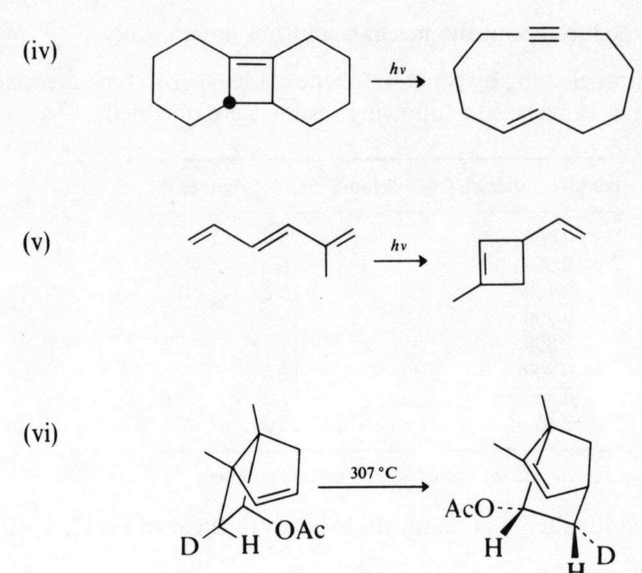

40. Show that, in benzene, the transitions $^1A_{1g} \longrightarrow {}^1B_{2u}$ and $^1A_{1g} \longrightarrow {}^1B_{1u}$ are symmetry-forbidden, and that the transition to the $^1E_{1u}$ state is allowed.

41. The $n \longrightarrow \pi^*$ transition in formaldehyde is symmetry-forbidden. Work out the symmetries of the vibrations that make the transition weakly allowed through vibronic coupling.

42. The diagrams below give the HOMO and LUMO for the benzyl cation ($PhCH_2^+$). By reference to the appropriate character table (C_{2v}) determine whether or not the lowest-energy $\pi \longrightarrow \pi^*$ transition is allowed, and, if it is, the direction of polarization.

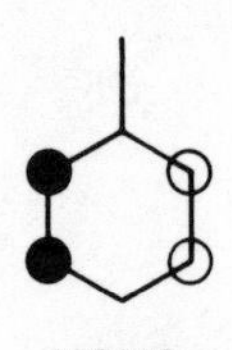

HOMO

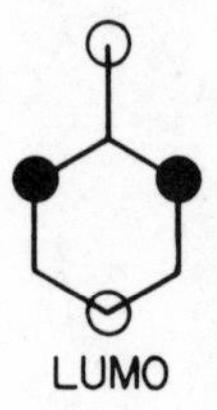

LUMO

Answers

1. Energy per mol = $Nh\nu = Nhc/\lambda$

   $\therefore$ 184·7 nm ≡ 648 kJ mol^{-1} ≡ 155 kcal mol^{-1}

   253·7 nm ≡ 472 kJ mol^{-1} ≡ 113 kcal mol^{-1}

   366·0 nm ≡ 327 kJ mol^{-1} ≡ 78·2 kcal mol^{-1}

2. 313 nm ≡ 382·5 kJ mol^{-1}

   Number of einsteins of radiation absorbed/second

   $= 6/382\,500 = 1{\cdot}57 \times 10^{-5}$

   $\therefore$ Number of moles of product formed/second

   $= 0{\cdot}470 \times 10^{-5}$

   $\therefore$ Time for 0·1 mole to be formed

   $= 0{\cdot}1/(0{\cdot}470 \times 10^{-5}) = 21\,300$ seconds

3. (i) O.D. $= \log(I_0/I) = \log(100/79{\cdot}4) = 0{\cdot}10$

   (ii) O.D. = 0·20 for 20 cm path length

   $\therefore$ $I_0/I = 1{\cdot}585$

   $\therefore$ 63·1 % is transmitted

   (iii) O.D. $= \varepsilon . c . l$

   $\therefore$ $\varepsilon = 1$ l mol^{-1} cm^{-1}

   (iv) No—ε is too low

4. From a re-plotted spectrum, $\int \varepsilon . d\nu \sim 8\text{–}10 \times 10^4$

   From equation (1.4),

   $f = 4{\cdot}3 \times 10^{-9} \int \varepsilon . d\nu$

   $\therefore$ $f \approx 4 \times 10^{-4}$

5. (i) $\int \varepsilon . d\nu \approx 10^3 \times 5000$

   $\therefore$ $f \approx 4.3 \times 10^{-9} \times 5 \times 10^6$ (equation 1.4)

   $= 0{\cdot}022$

   (ii) $\int \varepsilon . d\nu \approx 160 \times 4000$

   $\therefore$ $f = 4.3 \times 10^{-9} \times 64 \times 10^4$

   $= 0{\cdot}0028$

6. From the equations

$$A_{\text{ul}} = \frac{8\pi^2 v^2 e^2 f}{m_e c^2} = \frac{8\pi^2 e^2 f}{m_e c \lambda^2}$$

$$= \frac{8\pi^2 (4{\cdot}8 \times 10^{-10})^2}{(9{\cdot}1 \times 10^{-28})(3 \times 10^{10})(490 \times 10^{-7})^2}\ \text{s}^{-1}$$

$$= 2{\cdot}8 \times 10^8\ \text{s}^{-1}$$

$$\tau = A_{\text{ul}}^{-1} = 3{\cdot}6\ \text{ns}$$

7. Energy per pulse $= 200 \times 10^3 \times 10 \times 10^{-9}$ J

$= 0{\cdot}002$ J 5

Energy per photon $= hc/\lambda = \dfrac{6{\cdot}626 \times 10^{-34} \times 3{\cdot}0 \times 10^{10}}{337 \times 10^{-7}}$ J

$= 5.90 \times 10^{-19}$ J

∴ Number of photons $= 0{\cdot}002/(5.90 \times 10^{-19}) = 3{\cdot}4 \times 10^{15}$

8. Equation (2.22) gives

$$\ln (K^*/K) = \frac{h\,\Delta v}{kT} = \frac{hc(1/\lambda - 1/\lambda')}{kT}$$

$$= \frac{6{\cdot}626 \times 10^{-34} \times 3{\cdot}0 \times 10^{10} \times (1/475 - 1/550) \times 10^7}{1{\cdot}38 \times 10^{-23} \times 300}$$

$$= 13{\cdot}79$$

$$\therefore \quad \Delta(\text{p}K_a) = -13{\cdot}79/2{\cdot}303$$

$$= -5{\cdot}99$$

∴ pK_a (excited state) = 3·2

9. The method involves observation of the triplet–triplet absorption spectra in solutions of different pH (see G. Jackson and G. Porter, *Proc. Roy. Soc.*, **260A**, 13 (1961)).

10. In the excited state, contributions from resonance canonicals with charge on the ring carbon atoms play an important role. For the *o*- and *p*-substituted phenate anions there are canonicals (A, B) which make a smaller contribution because of the electronic effect of the methoxy group. This leads to the acidity of the excited state being lower for these isomers than for the *m*-isomer.

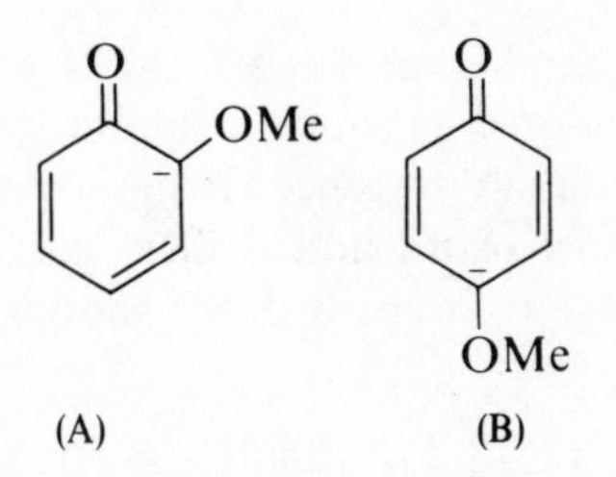

11. For benzophenone the long-wavelength band arises from an $n \longrightarrow \pi^*$ transition; for bis(dimethylamino)benzophenone the long-wavelength band is $\pi \longrightarrow \pi^*$.

12. The benzophenone phosphorescence is from a $^3(n, \pi^*)$ state associated largely with the C=O group; that of l-acetylnaphthalene is from a $^3(\pi, \pi^*)$ state associated with the whole π-system.

13. (i) Equation (3.15) gives $k_{isc} = \phi_{isc}/\tau_F$

$$\therefore \quad k_{isc} = 0{\cdot}50/3{\cdot}4 \times 10^{-8}\ s^{-1}$$
$$= 1{\cdot}5 \times 10^7\ s^{-1}$$

(ii) From p. 14, $\tau_F = \tau_0 \phi_F$

$$\therefore \quad \tau_0 = 3{\cdot}4 \times 10^{-8}/0{\cdot}14\ s$$
$$= 2{\cdot}4 \times 10^{-7}\ s$$

14. S_1 undergoes processes with rate constants k_F, k_{ic} and k_{isc}; S_2 undergoes processes with rate constants k'_F, k'_{ic} and k'_{isc}.

$\phi_F(S_1) = k_F \,.\, \tau; \quad \therefore \quad k_F = 0{\cdot}19/(9{\cdot}6 \times 10^{-8}) = 2{\cdot}0 \times 10^6\ s^{-1}$

$\phi_F(S_2) = k'_F \,.\, \tau' < 10^{-4}$

The first assumption we need to make is that $k'_F \geqslant k_F$; then

$$\tau' < 10^{-4}/(2{\cdot}0 \times 10^6)\ s$$

i.e.

$$\tau' < 5 \times 10^{-11}\ s$$

Now $\tau' = (k'_F + k'_{ic} + k'_{isc})^{-1}$, and we need to make the second assumption that internal conversion from S_2 to S_1 is the major contributor to this term. Then $k'_{ic} > 2 \times 10^{10}\ s^{-1}$

15. An external heavy-atom effect is operating (see p. 82), decreasing the quantum yield for fluorescence (20 000–27 000 cm^{-1} band) and increasing the quantum yield for phosphorescence (11 000–16 000 cm^{-1} band).

16. The mirror-symmetry relationship shows that the triplet state involved in delayed fluorescence is the same as the phosphorescent state.

17. The major factor is that the rate constant for ionization of the excited deuterated phenol (ArOD*) is less than that of the excited normal phenol (ArOH*).

18. The spectra suggest that higher vibrational levels of the electronically excited state undergo some chemical reaction (e.g. ring-opening) that competes effectively with fluorescence. If the experiments were repeated with increasing pressures of an added inert gas, vibrational relaxation should allow fluorescence to occur with the shorter wavelengths of excitation.

19. (i) No—the energy gap (T_1–S_0) is usually too large for organic molecules. (ii) Coupling will be greater for benzophenone because it has the smaller energy gap (S_1–T_1). This will result in a very much higher rate constant for intersystem crossing.

20. Substituents Me,OH,F increase k_f slightly and decrease k_{isc} with the result that ϕ_f is higher for these substituted naphthalenes. In the series of halonaphthalenes a heavy-atom effect operates, markedly increasing k_{isc} and k_p ($= \tau_p^{-1}$), and increasing ϕ_p. k_f is not affected, and so ϕ_f decreases considerably from fluoro- to iodo-naphthalene.

21. (i)
$$k_{isc} = \phi_{isc}/\tau_S$$
$$= 0{\cdot}97/(24 \times 10^{-9}) = 4{\cdot}0 \times 10^7\ s^{-1}$$

(ii)
$$\phi_f = 0{\cdot}03$$
$$\therefore\ k_f = 0{\cdot}03/(24 \times 10^{-9}) = 1{\cdot}2 \times 10^6\ s^{-1}$$

(iii) $k_p = \theta_p/\tau_T$, where θ_p is the quantum efficiency of phosphorescence (see p. 69).

$$\theta_p = \phi_p/\phi_{isc} = 0{\cdot}15/0{\cdot}97$$
$$\therefore\ k_p = \frac{0{\cdot}15}{0{\cdot}97 \times 0{\cdot}0015} = 103\ s^{-1}$$

(iv)
$$k_{NR} = \theta_{NR}/\tau_T$$
$$\theta_{NR} = \phi_{NR}/\phi_{isc} = (0{\cdot}97–0{\cdot}15)/0{\cdot}97$$
$$\therefore\ k_{NR} = \frac{0{\cdot}82}{0{\cdot}97} \times \frac{1}{0{\cdot}0015} = 560\ s^{-1}$$

22. *o*-Xylene forms a triplet eximer.

23. (i) The fluorescence of benzene monomer is quenched by ground-state benzene, and no excimer emission is seen. For pyrene, excimer fluorescence replaces monomer fluorescence as the concentration of pyrene increases (see p. 103), and the quantum yield for excimer fluorescence increases as the steady-state concentration of excimer increases.
(ii) Intersystem crossing ($S_1 \rightsquigarrow T_1$) is very much faster in the C=O compound, and fluorescence is therefore much weaker.

24. (i) $$\frac{d[D]}{dt} = k_D[E] = \frac{k_D}{(k_D + k_{-e})} \cdot k_e[^1A][A]$$

$$= \frac{k_D}{(k_D + k_{-e})} \cdot \frac{k_e[A]}{(k_e[A] + k_{isc} + k_{-1})} \cdot I$$

$$\therefore \quad \phi_D^{-1} = \left(1 + \frac{k_{-e}}{k_D}\right)\left(1 + \frac{(k_{isc} + k_{-1})}{k_e[A]}\right)$$

(ii) The process $E \rightarrow {}^1A + A$ has been neglected.

25. Yes, this would provide a spin-allowed two-step route for intersystem crossing.

26. The bands from 330–410 nm are the S_1 fluorescence spectrum of vinylcarbazole (mirror-symmetry relationship with the absorption spectrum). This emission is quenched by increasing concentrations of fumaronitrile and replaced by a broad, structureless band (λ_{max}550 nm) typical of exciplex emission. Charge-transfer within the exciplex gives rise to radical ions which are responsible for the ready polymerization of vinylcarbazole.

27. X is a triple exciplex. It must contain one molecule of A and two of D (concentration effect), and in the light of the solvent effect it must have an unsymmetrical structure (D.D.A)* rather than the symmetrical structure (D.A.D)* which would have a zero dipole moment.

28. R = rubrene, A = dimethylanthracene
$R + {}^1O_2 \rightarrow RO_2$, rate constant k_R
$A + {}^1O_2 \rightarrow AO_2$, rate constant k_A

$$-d[R]/dt = k_R[R][^1O_2]$$
$$-d[A]/dt = k_A[A][^1O_2]$$

$$\therefore \quad \frac{d[R]}{k_R[R]} = \frac{d[A]}{k_A[A]}$$

$$\therefore \quad k_A \ln[R] - k_R \ln[A] = k_A \ln[R]_O - k_R \ln[A]_O$$

where the subscript O refers to time zero.

$$\therefore \quad \frac{\ln([R]/[R]_O)}{\ln([A]/[A]_O)} = \frac{k_R}{k_A}$$

This is the expression required.
From the experimental results, this ratio is about 1·38.

29. For a discussion of quenching by oxygen, see section 4.5.3.

30. B = benzene, C = *cis*-butene, T = *trans*-butene

$B \xrightarrow{h\nu} {}^1B$, rate I

${}^1B \rightarrow {}^3B$, rate constant k_{isc}

${}^1B \rightarrow B$, rate constant k_{-1}

${}^3B \rightarrow B$, rate constant k_{-3}

${}^3B + C \rightarrow B + {}^3C/T$, rate constant k_q

${}^3C/T \rightarrow C$, rate constant k_C

${}^3C/T \rightarrow T$, rate constant k_T

$$\frac{d[T]}{dt} = k_T[{}^3C/T] = \frac{k_T}{(k_T + k_C)} k_q[C][{}^3B]$$

$$k_{isc}[{}^1B] = (k_{-3} + k_q[C])[{}^3B]$$

$$I = (k_{isc} + k_{-1})[{}^1B]$$

$$\therefore \quad \frac{d[T]}{dt} = \frac{k_T}{(k_T + k_C)} \cdot \frac{k_q[C]}{(k_{-3} + k_q[C])} \cdot \frac{k_{isc}}{(k_{isc} + k_{-1})} \cdot I$$

$$\phi_{isc} = \frac{k_{isc}}{(k_{isc} + k_{-1})}$$

$$\therefore \quad \phi_T^{-1} = (\phi_{isc} \,.\, F)^{-1}\left(1 + \frac{k_{-3}}{k_q[C]}\right)$$

where F is the branching ratio $k_T/(k_T + k_C)$.

A plot of ϕ_T^{-1} against $[C]^{-1}$ gives $(\phi_{isc} \,.\, F)^{-1}$ as the intercept, and F needs to be determined independently.

31. Mechanism (1):

$M \xrightarrow{h\nu} {}^1M$, rate I

${}^1M \rightarrow M$, rate constant k_{-1}

${}^1M \rightarrow {}^3M$, rate constant k_{isc}

${}^3M \rightarrow M$, rate constant k_{-3}

${}^3M + {}^3O_2 \rightarrow M + {}^1O_2$, rate constant k_q

${}^1O_2 \rightarrow {}^3O_2$, rate constant k_d

$M + {}^1O_2 \rightarrow MO_2$, rate constant k_r

$$\frac{d[MO_2]}{dt} = \frac{k_r[M]}{(k_r[M] + k_d)} \cdot \frac{k_q[{}^3O_2]}{(k_q[{}^3O_2] + k_{-3})} \cdot \frac{k_{isc} \,.\, I}{(k_{-1} + k_{isc})}$$

Mechanism (2):

The first four steps are as above, then:

${}^3M + {}^3O_2 \rightarrow MO_2$, rate constant k_r'

$$\frac{d[MO_2]}{dt} = \frac{k_r[{}^3O_2]}{(k_{-3} + k_r'[{}^3O_2])} \cdot \frac{k_{isc} \,.\, I}{(k_{-1} + k_{isc})}$$

The mechanisms can be distinguished by observing the variation of ϕ_{MO_2} with [M], provided that k_d is not negligible in comparison with k_r[M].

32. The *trans*-dimer is formed through the triplet state of (A), the *cis*-dimer through the singlet state.

33. The mechanisms are different for the two cycloalkenes. For 1-methylcyclopentene radical processes are occurring, involving hydrogen abstraction from the methyl group of CH_3OD. For 1-methylcyclohexene ionic processes are occurring, involving protonation by the —OD group of CH_3OD.

34. The reaction proceeds by way of a 1:1 adduct (C) which can be trapped by added dienophile, but only by a very powerful dienophile (see W. Hartmann *et al.*, *Tetrahedron Letters*, 883 and 3101 (1974); D. Bryce-Smith *et al.*, *ibid.*, 1627, (1975)).

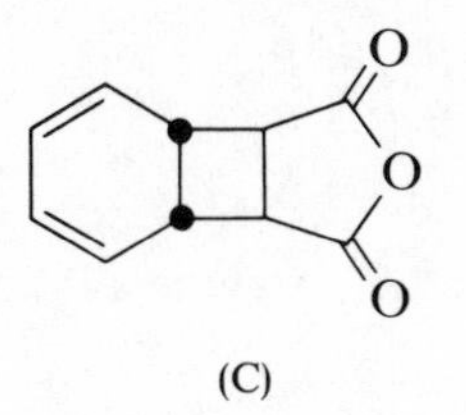

(C)

35. Equation (5.22) applies:

$$\frac{\phi^0}{\phi} = 1 + k_q\tau_0[Q]$$

$$\therefore \quad 100 = 1 + 10^{10} \times 0{\cdot}002[Q]$$

i.e. $[Q] \approx 5 \times 10^{-6}$ mol l^{-1}
The values of ϕ_p and τ_T are not required.

36. A plot of [aldehyde]$_0$/[aldehyde] against [penta-1,3-diene] is a straight line, and therefore only one excited state (triplet) is being quenched (see also p. 154). The enal is therefore formed from the triplet state, and the value of $k_q\tau_0$ is 20 l mol^{-1}.

37. A plot of [octanone]$_0$/[octanone] against [penta-1,3-diene] is *not* linear but rather of the form shown in Figure 5.7. Reaction must be occurring through two excited states of the ketone, probably the first excited singlet and triplet.

38. $$\frac{\phi}{0.22} = \frac{0{\cdot}0086}{0{\cdot}020}$$

$$\therefore \quad \phi = 0{\cdot}095$$

It is assumed that both ketones absorb the same fraction of the incident light, and that there is no competitive absorption by products nor quenching of the photoreaction by products.

39. (i) $(\pi^2 + \sigma^2)$
 (ii) $(\pi^2\text{s} + \sigma^2\text{s})$
 (iii) $(\pi^4 + \sigma^2)$
 (iv) $(\pi^2\text{s} + \sigma^2\text{s})$
 (v) $(\pi^2 + \pi^2)$
 (vi) $(\pi^2\text{s} + \sigma^2\text{a})$

40. The relevant direct products are:

(x, y axes)
$$A_{1g} \times E_{1u} \times B_{2u} = E_{2g}$$
$$A_{1g} \times E_{1u} \times B_{1u} = E_{2g}$$
$$A_{1g} \times E_{1u} \times E_{1u} = A_{1g} + \text{other terms}$$

(z axis)
$$A_{1g} \times A_{2u} \times B_{2u} = B_{1g}$$
$$A_{1g} \times A_{2u} \times B_{1u} = B_{2g}$$
$$A_{1g} \times A_{2u} \times E_{1u} = E_{1g}$$

Only the transition to the $^1E_{1u}$ state is allowed, polarized in the x and y directions.

41. The ground state of formaldehyde has symmetry A_1, the (n, π^*) excited state has symmetry A_2, ($O(2p_x)$ belongs to the B_1 irreducible representation, π^* to the B_2 irreducible representation, and $B_1 \times B_2 = A_2$). The cartesian coordinates have symmetries $B_1(x)$, $B_2(y)$, $A_1(z)$, and the direct products are:

$$A_1 \times B_1 \times A_2 = B_2$$
$$A_1 \times B_2 \times A_2 = B_1$$
$$A_1 \times A_1 \times A_2 = A_2$$

Vibrations of symmetries B_2, B_1 or A_2 will therefore make the transitions weakly allowed through vibronic coupling.

42. The HOMO belongs to the A_2 irreducible representation, and the LUMO to the B_2 irreducible representation (the xz plane contains the carbon atoms). The transition will therefore be allowed only in the x-direction, since the x axis has B_1 symmetry, and $A_2 \times B_1 \times B_2 = A_1$.

Appendix

Applications of Group Theory

Although desirable, it is not essential that those who use group theory should really understand its mathematical basis,[1] and in this appendix group theory is treated as a 'black box' for dealing with symmetry questions, a given input leading to a characteristic output.

A.1 SYMMETRY ELEMENTS AND SYMMETRY OPERATIONS

The symmetries of constructs such as molecules, wavefunctions, vibrations, Hamiltonian operators and so on can be classified by noting the symmetry *elements* (e.g., planes of symmetry) which they possess and determining whether they are symmetric or antisymmetric with respect to the corresponding symmetry operation. A system which remains unchanged by a symmetry *operation* (e.g., rotation or reflection) is said to be symmetric with respect to that operation. Conversely, a system changed into its inverse by the symmetry operation is designated as antisymmetric. For example, the ammonia molecule (A.1) is symmetric to rotations of 0°, 120° and 240° about the axis shown, but ψ_2 of *s-trans* butadiene (A.2) is antisymmetric to a rotation through 180°.

There are five independent symmetry operations:

(i) The identity operation, designated E, i.e. the operation 'do nothing', the effect of which is to leave the system unchanged.

(ii) Rotations about an axis through $(360/n)°$, designated C_n. Thus ammonia has a C_3 axis, *s-trans* butadiene a C_2 axis, and benzene (A.3) has a C_6 principal axis and a number of C'_2 axes perpendicular to the principal axis.

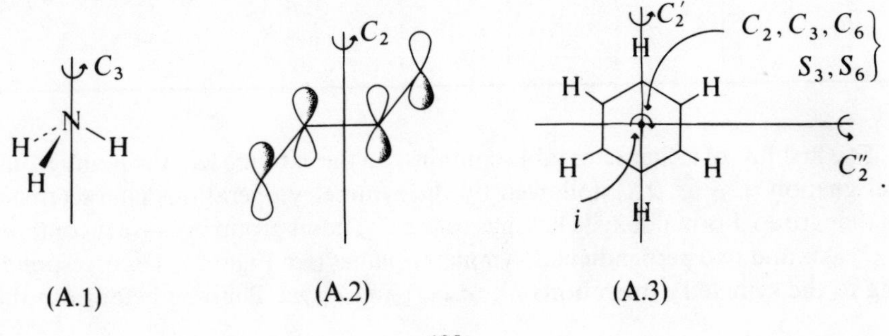

(A.1) (A.2) (A.3)

(iii) Reflections, either horizontal in the plane of the paper (σ_h), vertical to the plane of the paper (σ_v), or dihedral (σ_d) which bisect the angle between the two-fold axes (C'_2 and C''_2) perpendicular to the C_n principal axis. Benzene has all three classes of planes of symmetry.

(iv) Inversion through a centre of symmetry (i). A centre of symmetry is possessed by benzene and by staggered ethane (A.4).

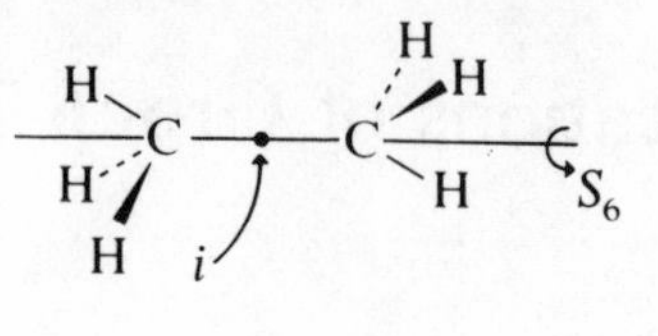

(A.4)

(v) Rotation–reflection—one or more repetitions of the sequence: rotation by $(360/n)°$ followed by reflection in a plane perpendicular to the rotation axis. This symmetry element is called an improper rotation and is given the symbol S_n. Staggered ethane has a S_6 axis.

Application of group theory involves selection of the appropriate group character table and use of the data enshrined therein.

A.2 CHARACTER TABLES

An arrangement of points or atoms in space may have certain symmetry elements, and a knowledge of these symmetry elements is enough to specify a particular *point group*, designated by a symbol such as C_{2v} or D_{6h}. The first step in any group theoretical analysis is to allocate the molecule under consideration to a particular point group. In order to illustrate the procedure, the character tables for the C_{2v} and D_{6h} point groups are reproduced below and opposite.

C_{2v}	E	C_2	$\sigma_v(xz)$	$\sigma'_v(yz)$	Co-ordinates and rotations	d-orbitals
A_1	1	1	1	1	z	x^2, y^2, z^2
A_2	1	1	−1	−1	R_z	xy
B_1	1	−1	1	−1	x, R_y	xz
B_2	1	−1	−1	1	y, R_x	yz

The first line of a character table contains, on the extreme left, the point group designation (C_{2v} or D_{6h}), followed by the symmetry operations characteristic of that group. Formaldehyde belongs to the C_{2v} point group because it contains a C_2 axis and two perpendicular symmetry planes (see Figure A.1), corresponding to the symmetry operations C_2, $\sigma_v(xz)$ and $\sigma'_v(yz)$. Benzene belongs to the

D_{6h}	E	$2C_6$	$2C_3$	C_2	$3C'_2$	$3C''_2$	i	$2S_3$	$2S_6$	σ_h	$3\sigma_d$	$3\sigma_v$	Co-ordinates and rotations	d-orbitals
A_{1g}	1	1	1	1	1	1	1	1	1	1	1	1		$x^2 + y^2, z^2$
A_{2g}	1	1	1	1	−1	−1	1	1	1	1	−1	−1	R_z	
B_{1g}	1	−1	1	−1	1	−1	1	−1	1	−1	1	−1		
B_{2g}	1	−1	1	−1	−1	1	1	−1	1	−1	−1	1		
E_{1g}	2	1	−1	−2	0	0	2	1	−1	−2	0	0	(R_x, R_y)	(xz, yz)
E_{2g}	2	−1	−1	2	0	0	2	−1	−1	2	0	0		$(x^2 - y^2, xy)$
A_{1u}	1	1	1	1	1	1	−1	−1	−1	−1	−1	−1		
A_{2u}	1	1	1	1	−1	−1	−1	−1	−1	−1	1	1	z	
B_{1u}	1	−1	1	−1	1	−1	−1	1	−1	1	−1	1		
B_{2u}	1	−1	1	−1	−1	1	−1	1	−1	1	1	−1		
E_{1u}	2	1	−1	−2	0	0	−2	−1	1	2	0	0	(x, y)	
E_{2u}	2	−1	−1	2	0	0	−2	1	1	−2	0	0		

D_{6h} point group because it contains the symmetry elements of that group (A.3). Thus a molecule can be assigned to a point group by writing down the symmetry elements or operations appropriate to the molecular framework and matching these against those displayed on the top line of the various character tables.

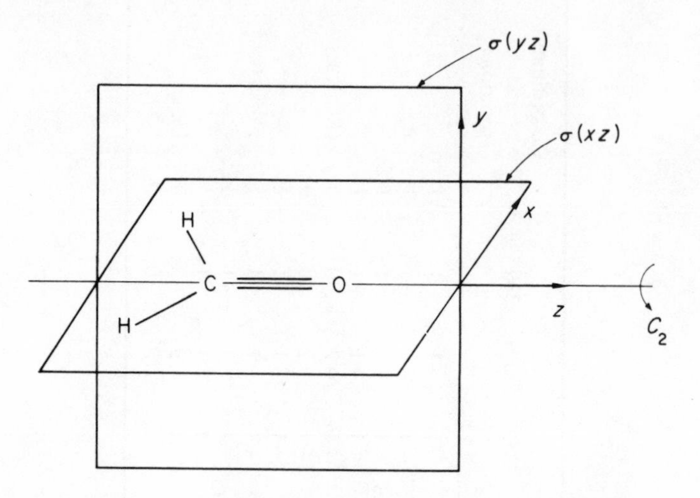

Figure A.1. Formaldehyde belongs to the C_{2v} point group

In the body of the table, the numbers 0, 1, 2 etc. are the characters of the irreducible representations of the point group, one representation per line and each labelled with the appropriate symmetry symbol (A_{1g}, B_2 etc.). What is really meant by the 'character of an irreducible representation' is best discovered from a standard text on group theory, but it is sufficient for our purposes to say that a set of matrices can be found, each corresponding to a single symmetry operation of the group, which combine in the same way as the symmetry operations. The simplest matrices of this sort are said to be irreducible representations of the group. The character, or trace, of a matrix is the sum of the elements constituting its main diagonal.

To illustrate this point, take the C_{2v} point group and the co-ordinate system of Figure A.2, and consider the way in which the vector (x_1, y_1, z_1) transforms under the symmetry operations of the group. If, under the symmetry operation $\sigma(xz)$, i.e. reflection in the xz plane, it is transformed into the vector (x_2, y_2, z_2), then the transformation can be formulated in terms of the equations (A.5).

$$
\begin{aligned}
x_2 &= x_1 + 0y_1 + 0z_1 \\
y_2 &= 0x_1 - y_1 + 0z_1 \\
z_2 &= 0x_1 + 0y_1 + z_1
\end{aligned}
\tag{A.5}
$$

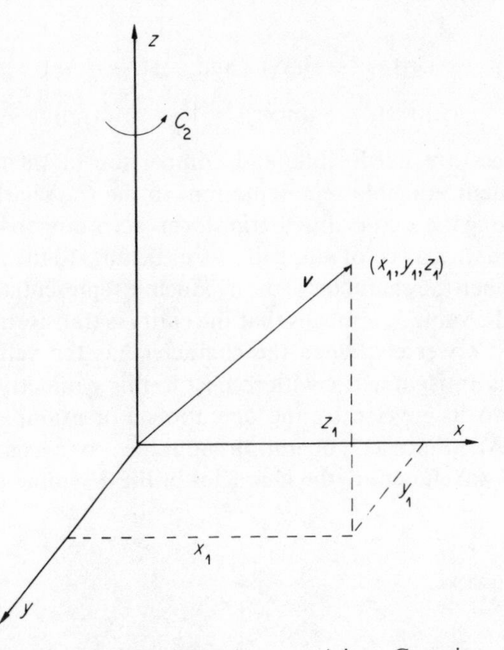

Figure A.2. The vector (x_1, y_1, z_1) in a Cartesian coordinate system

These equations can be written in matrix form (A.6).

$$\begin{pmatrix} x_2 \\ y_2 \\ z_2 \end{pmatrix} = \begin{pmatrix} 1 & 0 & 0 \\ 0 & -1 & 0 \\ 0 & 0 & 1 \end{pmatrix} \begin{pmatrix} x_1 \\ y_1 \\ z_1 \end{pmatrix} \tag{A.6}$$

The (3×3) matrix corresponds to the $\sigma(xz)$ symmetry operation; its character is $(1 - 1 + 1) = 1$, and we can write $\chi(\sigma(xz)) = 1$. Similarly, for the identity representation E,

$$\begin{pmatrix} x_2 \\ y_2 \\ z_2 \end{pmatrix} = \begin{pmatrix} 1 & 0 & 0 \\ 0 & 1 & 0 \\ 0 & 0 & 1 \end{pmatrix} \begin{pmatrix} x_1 \\ y_1 \\ z_1 \end{pmatrix} \quad \text{and} \quad \chi(E) = 3 \tag{A.7}$$

These matrices are not the simplest possible. By a rotation of the axes, the vector $\boldsymbol{v}$ can be brought into coincidence with, say, the x-axis (see Figure A.2). The matrices, with their characters, are now:
for C_2

$$\begin{pmatrix} x_2 \\ y_2 \\ z_2 \end{pmatrix} = \begin{pmatrix} -1 & 0 & 0 \\ 0 & 0 & 0 \\ 0 & 0 & 0 \end{pmatrix} \begin{pmatrix} x_1 \\ y_1 \\ z_1 \end{pmatrix}$$

or

$$(x_2) = (-1)(x_1) \quad \text{and} \quad \chi(C_2) = -1$$

$$\chi(E) = 1, \qquad \chi(\sigma(xz)) = 1, \qquad \chi(\sigma(yz)) = -1 \tag{A.8}$$

These matrices are irreducible, and comparison of their values with the characters of the irreducible representations in the C_{2v} character table shows that a vector along the x co-ordinate transforms according to the B_1 representation. In the same way, a vector along the z-axis belongs to the A_1 representation.

Notice that when the character of the irreducible representation of a symmetry operation has the value 1, it means that the entity is transformed into itself (i.e., is unchanged). Conversely, when the character has the value -1, it implies that the entity is antisymmetric with respect to this symmetry operation and is transformed into its inverse by the operation. For example, consider a $2p_x$ orbital (Figure A.3). Since a C_2 operation about the y- or z-axis inverts the phases or signs of this wavefunction, the character of these symmetry operations will be -1.

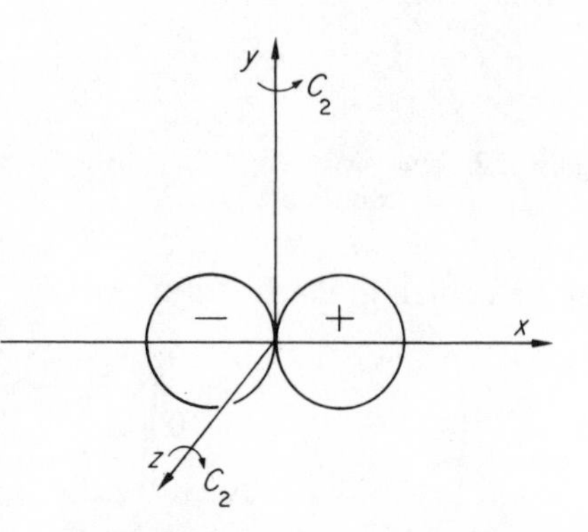

Figure A.3. A $2p_x$ orbital

In some symmetry groups there will be symmetry operations whose characters are 2 or 0.† These correspond to degenerate irreducible representations. To illustrate this point, consider the behaviour of the y co-ordinate axis under the symmetry operations of the D_{6h} point group (Figure A.4).

The y-axis is clearly antisymmetric to C_2 or i and symmetric to σ_h. However, the y-axis and the x-axis are neither symmetric nor antisymmetric to C_3 (rotation through 120°); instead, they are each converted into a linear combination of both in accordance with the equations (A.9).

$$\begin{aligned} x_2 &= x_1 \cos 120° - y_1 \sin 120° \\ y_2 &= x_1 \sin 120° + y_1 \cos 120° \end{aligned} \tag{A.9}$$

† Some character tables also contain trigonometric functions and complex numbers. These are rarely encountered and will not be discussed here.

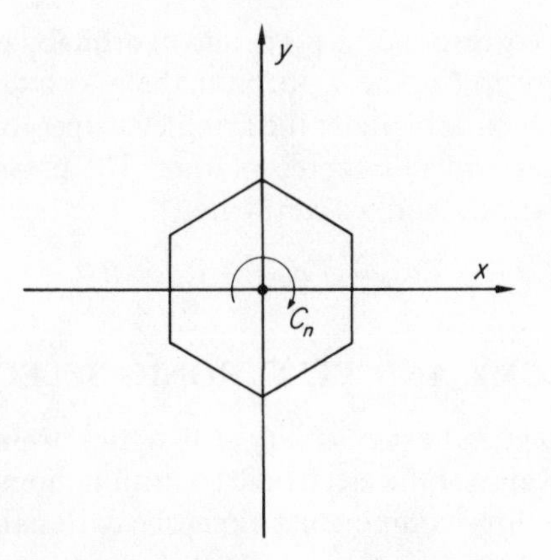

Figure A.4. Diagram for examining the behaviour of the y-axis under the symmetry operations of the D_{6h} point group

i.e.

$$\begin{pmatrix} x_2 \\ y_2 \end{pmatrix} = \begin{pmatrix} -\frac{1}{2} & -\frac{\sqrt{3}}{2} \\ \frac{\sqrt{3}}{2} & -\frac{1}{2} \end{pmatrix} \begin{pmatrix} x_1 \\ y_1 \end{pmatrix}$$

and

$$\chi(C_3) = -1 \tag{A.10}$$

Similarly, for the identity operation,

$$\begin{pmatrix} x_2 \\ y_2 \end{pmatrix} = \begin{pmatrix} 1 & 0 \\ 0 & 1 \end{pmatrix} \begin{pmatrix} x_1 \\ y_1 \end{pmatrix}$$

and

$$\chi(E) = 2 \tag{A.11}$$

Thus neither the x-axis nor the y-axis alone forms the basis for an irreducible representation of the group, but the pair taken together and written (x, y) generate (2×2) matrices which do transform properly under the group symmetry operations. The x and y co-ordinates, taken together, are said to form the basis of a *degenerate* representation of the group. Such irreducible representations are given the symmetry symbol E (to be distinguished from the identity operation also designated E).

Finally, the right-hand columns of the group character tables show to which irreducible representation the various mathematical functions or orbitals belong, such as rotations about the cartesian co-ordinate axes (R_x, R_y, R_z), the axes themselves, the corresponding p_x, p_y and p_z orbitals, and also the d-orbitals. Thus, in the point group C_{2v} the d_{yz}-orbital, the p_y-orbital and rotations about the x-axis all behave similarly under the symmetry operations of the group, and all belong to the B_2 irreducible representation. The upper case Greek gamma is used for representations, and we can write

$$\Gamma d_{yz} = \Gamma p_y = \Gamma R_x = B_2$$

A.3 GROUP THEORY AND ELECTRONIC SELECTION RULES

As discussed in Chapter 2, the probability of an optical transition is proportional to the square of the value of the electronic transition moment $\int \psi_i \boldsymbol{\mu} \psi_f \, d\tau$, which can be resolved into three components along the cartesian axes:

$$ETM = ETM_x + ETM_y + ETM_z$$

For a transition to be forbidden, all three component integrals must be zero. Alternatively, if just one of the integrals, say ETM_x, is non-zero, then the transition will be polarized along the x-axis. As discussed qualitatively on p. 17, any integral, taken over the whole of space, must necessarily be zero if the integrand is an odd (antisymmetric) function of the co-ordinates. In group theoretical language, unless the integrand belongs to the totally symmetric irreducible representation, the integral will be zero.

The group theoretical treatment of the problem is to assign the initial and final orbitals (ψ_i and ψ_f), between which the transition occurs, to particular irreducible representations ($\Gamma\psi_i$ and $\Gamma\psi_f$) of the point group of the molecule, and to do the same for the cartesian axes x, y and z (Γx, Γy and Γz). The *direct products* (e.g., $\Gamma\psi_i \times \Gamma x \times \Gamma\psi_f$) are then evaluated to find out to which irreducible representation these direct products belong. Unless the direct products contain the totally symmetric representation (A_1 or A_{1g}, the first line of the appropriate group character table), then the component electronic transition moments will be zero. Some examples will make the procedure clear.

(i) *The $n \rightarrow \pi^*$ Transition in Formaldehyde*

Formaldehyde belongs to the C_{2v} symmetry group. Using the co-ordinate system of Figure A.5, consider the symmetry of the O($2p_x$) orbital.

With respect to the symmetry operations of the group, E, C_2, $\sigma_v(xz)$ and $\sigma_v'(yz)$, the orbital is symmetric, antisymmetric, symmetric and antisymmetric respectively, i.e. the characters of the transforming matrices are 1, -1, 1 and -1. The O($2p_x$) orbital therefore belongs to the B_1 irreducible representation (see character table, p. 196).† Similarly, the π^* orbital is symmetric, antisymmetric, antisymmetric and symmetric with respect to these symmetry operations. The characters for these operations are therefore 1, -1, -1 and 1, so that this

† This result could have been obtained immediately by noting that p_x, p_y and p_z orbitals have the same symmetries as the x-, y- and z-axes respectively and by consulting the character table to discover the irreducible representation corresponding to the x-axis.

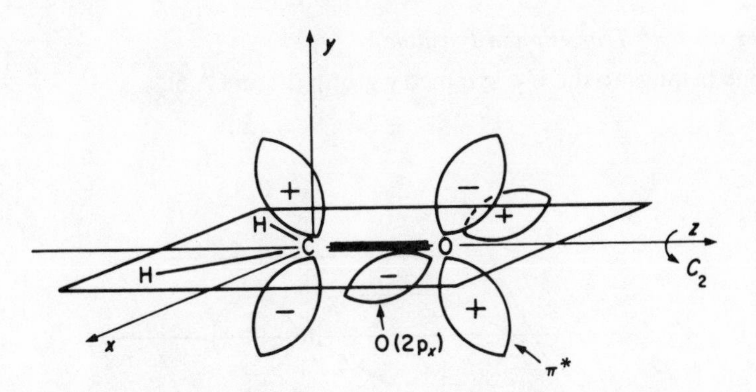

Figure A.5. O(2p$_x$) and π^* orbitals of formaldehyde

orbital belongs to the B_2 irreducible representation. The x-axis belongs to the B_1 representation. We now have to discover whether the direct product $B_1 \times B_1 \times B_2$ contains the totally symmetric representation A_1. This is done by multiplying together, for each symmetry operation in turn, the characters for those operations in each of the irreducible representations, as in the following table.

Symmetry operation	E	C_2	$\sigma_v(xz)$	$\sigma_v'(yz)$
Characters of direct product $B_1 \times B_1 \times B_2$	$1 \times 1 \times 1$ $=1$	$-1 \times -1 \times -1$ $=-1$	$1 \times 1 \times -1$ $=-1$	$-1 \times -1 \times 1$ $=1$
Characters of B_2 irreducible representation	1	−1	−1	1

The direct product $B_1 \times B_1 \times B_2$ is seen to generate the irreducible representation B_2. Since this is not the totally symmetric representation A_1, the integrand is an odd function of the co-ordinates, and $ETM_x = 0$.

Similarly,

$$\Gamma ETM_y = B_1 \times B_2 \times B_2 = B_1$$

and

$$\Gamma ETM_z = B_1 \times A_1 \times B_2 = A_2$$

Hence all three components of the electronic transition moment are zero, and the $n \rightarrow \pi^*$ transition in formaldehyde is predicted to be symmetry-forbidden.

(ii) *The n→ π* Transition in Pyridine*

Pyridine belongs to the C_{2v} symmetry group (Figure A.6).

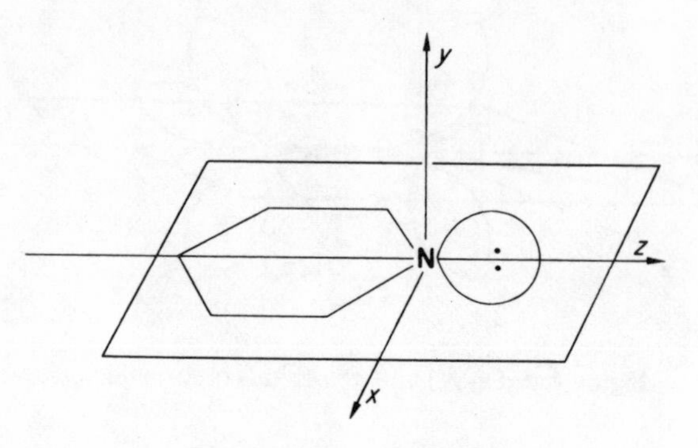

Figure A.6. Pyridine

The non-bonding orbital on nitrogen belongs to the A_1 irreducible representation. The lowest energy unfilled orbital of pyridine has the form of Figure A.7, and it belongs to the irreducible representation B_2.

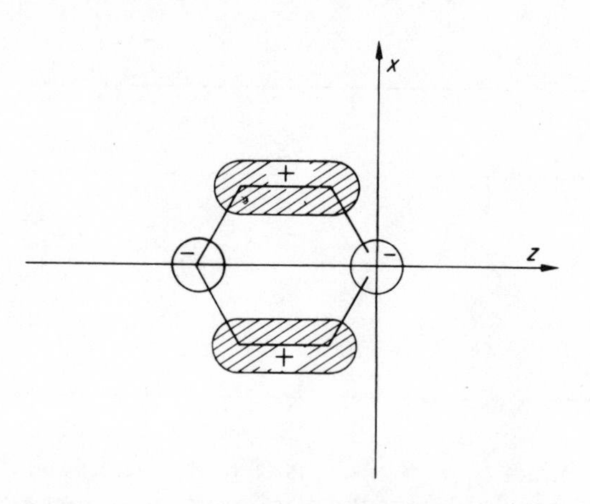

Figure A.7. The lowest unfilled (π^*) orbital of pyridine

The integrand of ETM_y belongs to the representation $A_1 \times B_2 \times B_2 = A_1$, and the transition is allowed if polarized along the y-axis.

(iii) *The Excited States of Benzene*

The molecular orbitals of benzene take the form given in Figure A.8, and in the D_{6h} point group belong to the irreducible representations shown (it is conventional to give orbitals lower case symbols).

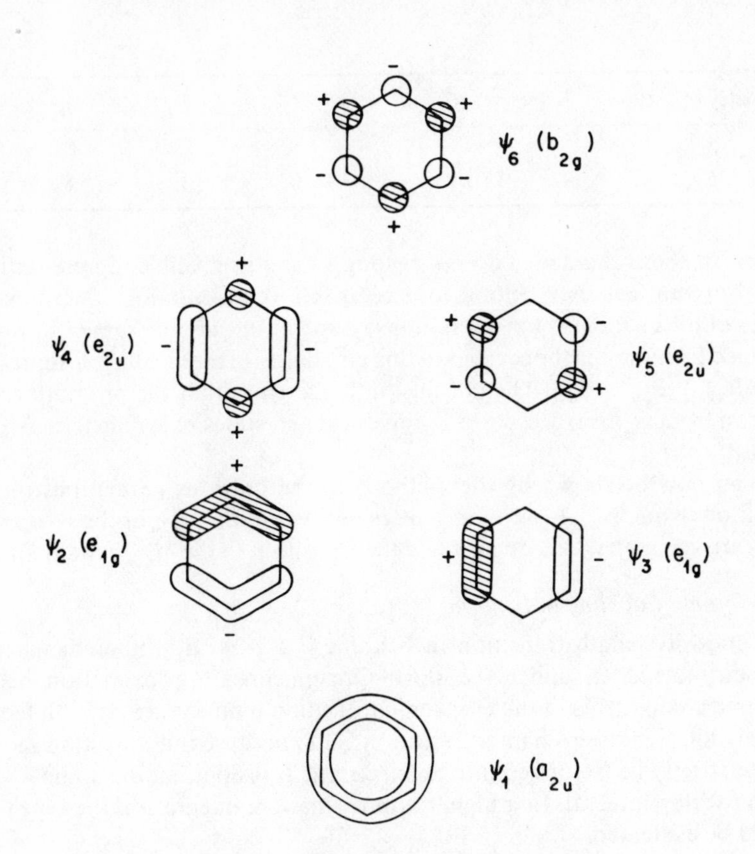

Figure A.8. Molecular orbitals of benzene

The lowest excited states of benzene have the electron configuration $(a_{2u})^2(e_{1g})^3(e_{2u})$, and in order to find the symmetries of these states it is necessary to calculate the direct product

$$
\begin{aligned}
& A_{2u} \times A_{2u} \times E_{1g} \times E_{1g} \times E_{1g} \times E_{2u} \\
&= (A_{2u} \times A_{2u}) \times (E_{1g} \times E_{1g}) \times E_{1g} \times E_{2u} \\
&= A_{1g} \times A_{1g} \times E_{1g} \times E_{2u}
\end{aligned}
$$

which reduces† to

$$E_{1g} \times E_{2u}$$

For symmetry operations of this point group the elements of this direct product are set out in the following table.

† Note that the direct product of two *identical* non-degenerate irreducible representations always contains the totally symmetric irreducible representation, and also that the totally symmetric irreducible representation can be struck out from any direct product containing it since its presence is without effect.

Symmetry operation	E	C_2	C_3	C_6	C_2'	C_2''	i	S_3	S_6	σ_h	σ_d	σ_v
Characters of $E_{1g} \times E_{2u}$	4	−1	1	−4	0	0	−4	1	−1	4	0	0

Clearly, these characters do not belong to any irreducible representation of the group; in fact they belong to a reducible representation. The numerical values of the characters for the symmetry operations are the same as would be obtained by summing the corresponding characters of the irreducible representations $B_{1u} + B_{2u} + E_{1u}$. The conclusion is therefore that the promotion of an electron from ψ_2 or ψ_3 to ψ_4 or ψ_5 gives excited states of symmetries B_{1u}, B_{2u} and E_{1u}.

It can now be shown by the methods of the previous paragraphs that the transitions from the $^1A_{1g}$ ground state of benzene to the $^1B_{2u}$ or the $^1B_{1u}$ excited states are symmetry-forbidden, while the transition to the $^1E_{1u}$ state is allowed.

(iv) *Vibronic Coupling in Benzene*

The long-wavelength transition in benzene ($^1A_{1g} \rightarrow {}^1B_{2u}$) is predicted to be symmetry-forbidden under the Born–Oppenheimer approximation because the three components of the electronic transition moment are zero. In fact it is 'slightly allowed' (see Chapter 2, section 2.1.8) because the transition moment cannot strictly be factorized into an electronic transition moment and a vibrational overlap integral. In a higher approximation, integrals of the type (A.12) should be evaluated,

$$\int \psi_{i,m} \mu \psi_{f,n} \, d\tau \tag{A.12}$$

where $\psi_{i,m}$ and $\psi_{f,n}$ are the vibronic wavefunctions corresponding to the *mth* vibrational level of the initial state and the *n*th vibrational level of the final state respectively.

In group theory terms, the symmetry of a given vibronic level can be obtained (i.e., assignment to a particular irreducible representation can be made) by taking the direct product of the irreducible representations of the electronic and vibrational components,

$$\text{i.e.} \quad \Gamma\psi_{i,m} = \Gamma\psi_i \times \Gamma\theta_m$$

Reverting to the benzene case, suppose that the initial state is the $v'' = 0$ level of the ground state. Then

$$\psi_{i,m} = {}^1A_{1g}(v'' = 0)$$

If a molecule is in its vibrational ground state, then its vibrational wavefunction is totally symmetric and belongs to the A_{1g} representation. Thus for the initial state the irreducible representation can be written as

$$\Gamma^1A_{1g}(v'' = 0) = A_{1g} \times A_{1g} = A_{1g}$$

The irreducible representations corresponding to the dipole moment operator $\boldsymbol{\mu}$ are the same as those of the three cartesian co-ordinates and are E_{1u} or A_{2u} (from the D_{6h} character table). The final state will belong to the representation

$$B_{2u} \times \Gamma\theta_n$$

The integral of equation (A.12) will now belong to the representation

$$\Gamma\psi_{i,m} \times \Gamma\mu \times \Gamma\psi_{f,n} = A_{1g} \times E_{1u} \times B_{2u} \times \Gamma\theta_n \tag{A.13}$$

or

$$A_{1g} \times A_{2u} \times B_{2u} \times \Gamma\theta_n \tag{A.14}$$

If the integral (A.12) is to have a non-zero value, then one of the expressions (A.13) or (A.14) must contain the totally symmetric irreducible representation A_{1g}. Taking expression (A.13), the direct product $\mathrm{A}_{1g} \times E_{1u} \times B_{2u}$ is readily verified as being equivalent to E_{2g}. Therefore the expression (A.13) will contain the A_{1g} representation if

$$\Gamma\theta_n = E_{2g}$$

Similarly for the expression (A.14)

$$A_{1g} \times A_{2u} \times B_{2u} = B_{1g}$$

and $\Gamma\theta_n$ must belong to the irreducible representation B_{1g}.

The conclusion is that the ${}^1A_{1g} \rightarrow {}^1B_{2u}$ transition in benzene will become 'slightly allowed' by vibronic coupling involving vibrations of the benzene molecule which have the B_{1g} or E_{2g} symmetry.

REFERENCES

1. An excellent introduction is F. A. Cotton, *The Chemical Applications of Group Theory*, Interscience, New York (1963); see also D. Schonland, *Molecular Symmetry*, Van Nostrand, London (1965). These books contain group character tables.

Index